SCIENCEBITES

SCIENCEBITES

A fresh take on commonly used terms in science

Gerard Jagers op Akkerhuis

EAN: 9789086863365
e-EAN: 9789086868872
ISBN: 978-90-8686-336-5
e-ISBN: 978-90-8686-887-2
DOI: 10.3920/978-90-8686-887-2

First published, 2019

©Wageningen Academic Publishers
the Netherlands, 2019

Wageningen Academic Publishers
P.O. Box 220, 6700 AE Wageningen
the Netherlands
www.WageningenAcademic.com
copyright@WageningenAcademic.com

Table of contents

Preface

Do you enjoy taking a step back and observing the world, and science, from a philosophically inspired perspective? Then the SCIENCEBITES in this book will offer challenging food for thought.

The name SCIENCEBITES refers to small 'bites' that can be nibbled, or chewed on, in between other activities. Every SCIENCEBITE offers a fresh take on a fundamental science term, such as life, evolution, organism, science, and big history.

The first SCIENCEBITE introduces a recent theory for analyzing complexity, called the operator theory. This theory acts as a context for most of the SCIENCEBITES in this book. Reading the first SCIENCEBITE will certainly help you understand the other SCIENCEBITES. Apart from Chapter 1, the SCIENCEBITES can be read in random order, because relevant information is always offered locally.

Acknowledgements

These SCIENCEBITES could not have been written without the help of colleagues and friends. In particular I thank Bart Gremmen for discussions during the writing of this book, and for sharing his thoughts while proofreading the final draft. I also enjoyed numerous thought-exchanges with Josette Jacobs, Henk van den Belt, and Leon Pijnenburg.

The lectures of Josette Jacobs kindled my interest in science criteria. She challenges her students to identify more than 12 criteria for scientific activity. The chapter on science criteria also profited from discussions with Bethany Sollereder and Bart Voorzanger.

I like to give special thanks to Herman Philipse for commenting on a selection of philosophically oriented chapters. His suggestions have helped improving the content and clarity of these texts. Several lunch meetings with Jelle Zandveld offered an opportunity to evaluate the educational value of the smallest pattern of Darwinian evolution, its extension and its generalization. I am grateful to Peter Roessingh, Marnix van Meer, Peter Schippers, and Diedert Spijkerboer for past and recent discussions about several topics that found their way into the SCIENCEBITES. Henk Barendrecht introduced me to the work of Carlo Rovelli ('Seven brief lessons on physics'). Rovelli's book 'The Order of Time' inspired me to writing a SCIENCEBITE about time. I thank Atanu Chatterjee for commenting on the chapter about the least action principle. Jan Meint Greben helped me by sharing his ideas about the physics of fundamental particles. The SCIENCEBITES about evolution and big history have profited from discussions with Bouwine Bergsma, Koert van Mensvoort, and Esther Quaedackers.

The text in this book aims at offering a condensed and integrated picture of the insights gained from the literature and from discussions with colleagues. Any misinterpretations are the sole responsibility of the author.

I thank my wife Florentine Jagers op Akkerhuis-Lammes for discussions about the content of some chapters. Florentine and Bart Gremmen helped me with the search for a title for this book.

Dr. Dr. Gerard Jagers op Akkerhuis

1. A remake of the *scala naturae*

Organizing a complex world. With every investigation of the world around us, we increase our knowledge. To handle all the information, and keep it accessible, it must constantly be organized, for example with the help of theoretical models.

The more knowledge becomes available, the more comprehensive and general our models of the world can become. One of the oldest general models, that has been very influential through the ages, dates back to Plato, and is called 'the ladder of nature', the Latin naming of which is the *scala naturae*. Others refer to this ladder as the 'great chain of being'. This chain classifies entities in order of increasing 'perfection'. At the bottom of the *scala* one finds clay and sand, then crystals, plants, animals, humans, and gods.

While offering a 'grand scheme', the *scala naturae* is not very popular amongst scientists these days. It is not viewed as a scientific model, but as an allegory. From a scientific perspective the *scala* is plagued by several problems. The first problem is that the ranking starts with physical objects, and continues towards supranatural entities (gods), while the focus of science is the natural world. A second obstacle is that the *scala* offers no causal explanation of how the entities in the hierarchy came to be. Thirdly, it is considered problematic that the *scala* offers no stringent definitions for the entities on the ladder. Ideally, such definitions must indicate what kind of organization is typical for every kind of entity.

It is likely that because of the above three problems scientists are reluctant to view the classical *scala* as a scientific approach. At the same time, however, criticism of the *scala* can also be viewed as inspiration for new theoretical development.

A remake. By rejecting the *scala naturae* because of obvious drawbacks, one risks throwing the baby – a grand scheme for hierarchical order – out with the bathwater. The alternative that I pursue in my work, is to improve the *scala* in such a way, that the result fits modern scientific requirements. Any initiative to improve the *scala* must start with a study of existing literature on hierarchical rankings. In this context I like to refer to the foundational works of Herbert Simon, Juan Alvarez de Lorenzana, James Miller, Eörs Szathmáry, John Maynard Smith and others.

Despite that the existing literature offers various mechanisms for creating hierarchical rankings, it intrigued me that all these sources still seem to struggle to avoid the same weaknesses as the classical *scala* suffers from. To clarify this point, the following hierarchy serves as an example:

cell → organ → organism → population (an arrow indicates a step towards a next level)

When analyzing this hierarchy, I will focus on the same three aspects that are viewed as drawbacks of the classical *scala*: (1) the inclusion of both physical and non-physical entities; (2) rules that don't offer causal explanations for transitions; and (3) unclarity about the typical properties defining an entity at any one level.

Kinds of entities. When discussing the *scala* I focused above on the difference in kind between natural entities and deities. Natural entities are physical things. As far as scientific investigations have been able to demonstrate, the deities are strictly associated with ideas in the minds of people. In this sense, deities are mental entities.

A similar shift from physical to non-physical/mental entities can also be observed in the above hierarchy from cell to population. Terms such as a cell, an organ, and a specific organism can be viewed as having physical entities as representations. In contrast, the term population refers to a 'group' of organisms. While the members of a group exist physically and individually, the fact that they can be viewed as a group represents a mental connection between a willful selection of entities. However real such connection may seem in our thinking, this does not make a group a physical thing in the field. While the organisms of the group exist physically, the group exists only in the mind. In line with this, a population is not a physical entity, but a mental one. This means that the above ranking changes in kind from physical entities to mental entities. For this reason, the entities in the ranking are not all of the same logical kind.

Causal explanations. The second discussion point is that the rules governing the steps from a cell to an organ, to an organism, and to a population, leave room for interpretational variability. One interpretation could be that 'X is a part of Y'. When looking at the hierarchy in this way, cells are parts of an organ, organs are parts of an organism, and organisms are parts of a population. Interestingly, a cell is not a part of an organ in the same way as an organism is part of a population. A cell is a physical part of a physical organ. And an organ is a physical part of a physical organism. But an organism, in spite of it being a physical entity, cannot physically be a part of the *mental* grouping known as a 'population', simply because concepts – like a population – have no physical parts. Groups have theoretical 'parts', called members. This leads to the conclusion that the 'is a part of' rule cannot be applied in the same way to every step in the hierarchy.

Unclarity about typical properties. Thirdly, the hierarchy is not explicit about the properties that characterize a cell as a cell, an organ as an organ, or an organism as an organism. Discussions of such properties are a philosophical minefield, because they are frequently viewed as leading towards 'essential' properties of entities, which properties are seen as being associated with Plato's 'idealism'. Idealists claims that for every entity there exists an eternal idea, an essence, that defines its properties.

Possibly because things are no longer thought to be static, but instead are thought to be the products of an ever changing, dynamic and evolving world, Plato's essences are not very popular amongst philosophers nowadays. Of course, one can use essential properties also in a non-Platonic way, implying that one searches for criteria that capture a kind of 'essence' of any and all entities that belong to a specific class, and not to any other class. Such criteria are known as being 'necessary and sufficient'.

Towards a new hierarchy theory. In the autumn of 1993 I worked on a project that brought to the surface the above challenges. As a contribution to the field, a new approach to hierarchy in nature was developed that makes use of two criteria for a hierarchical ranking:

1. The ranking must include objects of the same general kind (e.g. 'physical unit').

2. The ranking must follow one uniform rule that can be applied to every transition from one level to the next. As such a rule one can think for example of: 'Objects of kind A must interact in a specific way to form objects of a more complex kind B'.

When applying these two rules, I discovered that there is not just one way of ranking entities in a hierarchy, as in the above example, but that it is possible to imagine three distinct, and independently applicable kinds of hierarchy (Figure 1.1):

- The first kind of ranking focuses on the parts in singular objects, the parts in these parts, the parts in these parts, etc. This particular ranking can be viewed as pointing 'inward' at elements inside a 'whole' entity. Any inward ranking starts for example with an animal, and subsequently focuses on the limbs, the organs, the tissues and the cells in it. Any inward ranking depends on the point of view.

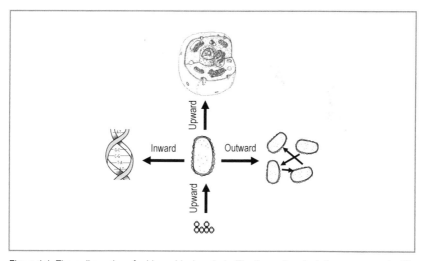

Figure 1.1. Three dimensions for hierarchical analysis. The 'inward' analysis focuses on parts of/in singular objects. The 'upward' analysis focuses on how singular objects in interaction create more complex, next level singular objects. The 'outward' analysis focuses on the many entities that can form when singular objects interact. The inward and outward analysis apply to every entity along the upward dimension.

- The second kind of ranking focuses selectively on singular objects and ranks these in order of decreasing or increasing complexity. With the current insights about nuclear physics, it is likely that – at least for the time being – quarks can be used as the foundation of this ranking. Starting with quarks one can construct more and more complex singular objects in an 'upward' way. The resulting upward ranking is not merely a classification but is the result of a causal chain of construction events. The classes of the upward ranking have instantiations of the following kinds: hadrons, atoms, molecules, cells, 'eukaryotic' cells and multicellular animals. If a singular object loses relevant organization, it moves one or more steps down the ranking.

- The third kind of ranking focuses on how we as humans group singular objects. Objects are grouped according to different kind of relationships/interactions. This ranking can be viewed as pointing 'outward', from singular objects to entities that are the product of interactions between many singular objects. I refer to such entities as 'interaction systems'. Any ranking of levels of organization in interaction systems depends on the point of view.

One can identify two major kinds of interaction systems (Figure 1.2). The first kind is called a 'compound object'. The criterion for a compound object is that it consists of singular objects and/or compound objects that are more strongly attached to each other than to the environment. Instances of the class of compound objects take the form of physically united, more or less organized 'lumps' of material, such as a stone, a car, and a planet.

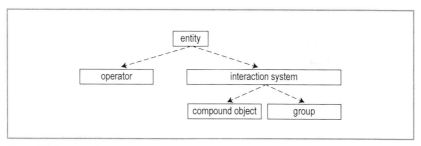

Figure 1.2. Classes of entities that form the top-levels of the terminology of the operator theory.

The second kind is called a 'group'. When using the term group here, I refer to a mental set that brings together a number of terms, each term referring to an individually countable, physical entity (if the objects would be attached, they would form a compound object). A group is a mental/conceptual entity. Examples of groups are: a population, a community, an ecosystem, a city, a country, and a football team.

Of the above three rankings (inward, upward, and outward), the upward ranking is special. The reason is that it allows one to single out a highly specific selection of events from the large possibility of space for interactions between objects. This selection describes how small singular objects aggregate in nature and form a new, and more complex, singular object, after which the story repeats. Every time the story repeats, the event implies a stepwise increase in organizational complexity of the entities involved. Step by step one can observe the formation of a series of entities with increasingly complex organization. This series offers a backbone for the analysis of hierarchy in nature.

Operator theory. The upward ranking was named 'the operator hierarchy'. Every singular object that, at any one level, fits the logic of the operator hierarchy is named an 'operator'. The field of theory related to the construction of the operator hierarchy, and its application, is named the operator theory.

I have chosen the name 'operator' in the operator theory in relation to historical publications of Conrad Hal Waddington, who used the term for the bodily aspect of organisms, which he contrasts with the genetic coding. Other uses of the term operator can be found in mathematics, and in the communication business. All the operators in the operator theory can be viewed as 'agents' in individual-based simulation models, in which e.g. individual atoms, molecules or organism interact on the basis of their physical and behavioral properties. However, not every entity that is called an agent in a model is an operator.

Foundation of the operator theory. To construct a stringent ranking, the operator hierarchy makes use of three criteria that define the kind of entities residing at a specific level of organization:

1. The least complicated opportunity to form a new kind of *process loop* (a functional requirement).
2. The least complicated opportunity to form a new kind of *containing layer* (a structural requirement).
3. The *mutual dependency* of the process loop and the containing layer.

As an example of how the above three criteria can be applied, one can think of a bacterial cell (Figure 1.3). The process loop inside a bacterium consists of a group of catalytic molecules that each produce one or more other catalysts from externally acquired resources. As soon as every catalytic molecule in the group has performed its catalytic action, all the catalytic molecules are re-produced. In the meantime other products are formed that are integrated into the membrane. The membrane forms a containing layer, an interface. In this way, the autocatalytic reactions produce molecules that become part of the membrane, while the membrane mediates the set of autocatalytic molecules. Through this interaction, the functioning of the membrane and the autocatalytic molecules obligatorily depend on each other.

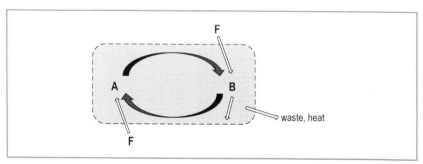

Figure 1.3. The bacterial cell as an example of the tree criteria involved in 'dual closure'. A dashed line indicates the containing layer (the structural closure) of the cell membrane. A black arrow indicates the process cycle (functional closure) representing a loop of catalytic reactions that re-produces every catalyst involved (depicted as a smallest possible set of catalysts A and B). The catalytic loop and the membrane depend on each other's existence. F = 'food' molecules. A white arrow indicates a flow of substance.

Dual closure. Of the above three criteria, the process loop, and the containing layer are the most paradigmatic. Because of a focus on paradigmatic organization and minimalism, and because the operator theory refers to a loop and to a containing layer as a 'closure', the cell is said to have 'dual closure'. The name 'dual closure' does not imply, however, that only two closures are involved. After all, the above listing requires three closures, the third closure describing that the process loop and containing layer must be mutually dependent. In fact, one could take into account additional aspects of the operator theory, and use four, five, or more criteria. Because such additional information is implied, the term dual closure does not literally cover its content. The name dual closure is used deliberately, to allow a focus on a minimalistic set of paradigmatic criteria.

Ranking dual closure steps: the operators. By stacking one dual closure event on top of the other, one obtains a causal hierarchy that includes increasingly complex kinds of operators. Starting with fundamental particles, e.g. quarks, the following operators have formed: the hadron, the atom, the molecule and the cell. From the cell, one can either follow a line towards the multicellular (bacterial multicellularity as found in blue-green algae), or one can follow a line that continues towards the endosymbiont cell, the endosymbiont multicellular, and the neural network organism.

Biologists historically refer to the endosymbiont cell as a cell with a nucleus/karyos, and for this reason as a 'eukaryotic cell'. However, in many cases the nucleus of the cell dissolves during cell division. What remains is a cell without a nucleus, which in a strict sense is not a eukaryote cell. Meanwhile, the endosymbionts, in the form of e.g. mitochondria and/or chloroplasts, remain present during the entire process of cell division. For this reason I consider the term 'endosymbiont cell' (a cell with inhabiting endosymbionts) more generally applicable in the context of the operator theory.

New possibilities. The above ranking of the operators adds a new and fundamental dimension to discussions about unity and hierarchy. Suddenly it becomes possible to define, in a stringent way, a broad range of differently sized 'elementary particles' called operators. While being elementary at any level, the operators are not 'fundamental' in the way quarks seem to be. Another way of looking at the operators is to view them as differently sized LEGO bricks in the organization of nature. Using combinations of smaller and larger LEGO bricks, one can build LEGO cities, LEGO towns, LEGO cars, etc.

Towards simplicity. The operators, and their hierarchical ranking, offer a lean context for re-evaluating, and in some cases redefining, current definitions of scientific concepts. When revisiting the descriptions of much-used science concepts, Ockham's razor, in its current interpretation, offers guidance. It says: make things as simple as possible.

With the aim of increasing simplicity I consider it valuable to continuously work on scientific communication by reducing 'conceptual friction'. Conceptual friction occurs when a definition is 'leaky', because it does not include all the desired instances and/or does not exclude all the non-desired instances. If a definition with a certain amount of conceptual friction can be replaced by a definition with less conceptual friction this can be viewed as a step forward in the simplicity and conceptual clarity of a theory/philosophy.

A theoretical example. As an example of the practical relevancy of the operator theory for e.g. biology, I compare the concept of multicellularity with that of pluricellularity. Before the operator theory, any grouping of cells was generally referred to as multicellular. With the help of the operator theory, however, one can make a clear distinction between pluricellularity, a concept which already exists in the literature, and multicellularity.

A pluricellular group consists of cells that are merely attached. Examples of pluricellular groups are e.g. the slug of a slime mold, the early human embryo, and a lichen.

When it comes to defining multicellularity, the operator theory and its criterion of dual closure invite one to look beyond the mere attachment of cells, and to focus on two criteria: (1) the presence of plasma connections between cells that allow the connected cells to, in a recursive way, influence each other's functioning (functional closure); and (2) the presence of a membrane surrounding a shared plasma (structural closure). Inside a multicellular organism, every cell of the tissues of the body (blood cells require a special discussion) will minimally be connected with one other cell in a way that allows for dual closure. The plasma connections offer a context for discussing the maintenance of all the connected cells as a unity.

A practical example. While knowing whether a unit is pluricellular or multicellular has theoretical importance, the question can be asked whether or not such knowledge also has practical relevancy? One field of practical relevancy is Darwinian evolution. If one speaks about multicellular organisms *sensu* the operator theory, descent and genetics are in line with each other. The reason is that multicellularity starts with a single cell from which the offspring descend clonally (a multicellular bud, and a chimera require some extra explanation as 'derived' properties).

In pluricellular groups, however, such alignment fails, because cells of different genetic origin can gather to form a group. The resulting group of cells must not be clonal. As a consequence, there will always be the possibility of, and a reward for, competition between genetic strains. In such cases, descent and genetics are not aligned.

The observation that pluricellular groups and multicellular organisms require a different approach when it comes to fitting their units into a theory of evolution, indicates the practical relevance of distinguishing these two kinds of organization.

Unexpected bonuses. Now that the operator hierarchy is available as a scientific tool, there is an interesting bonus because the operator theory offers a new perspective on a broad range of venerable concepts, such as organism, life/death, evolution, etc.

In addition, the operator theory has several philosophical implications. One implication is minimalism, in the sense that the focus lays on minimally required criteria. The second is generality, because the closure criteria are applicable throughout all levels of the operator hierarchy, from fundamental particles to conscious organisms. The third is necessity and sufficiency, which implies that the criteria used for the operator theory focus on the inclusion of every relevant example and the exclusion of every non-relevant example. The operator theory also contributes to philosophical discussions about what are 'individual entities', and what things comply with the concept of 'natural kinds'.

Science concepts as tools. Leveraged by the new theoretical possibilities of the operator theory, notably the innovative concept of dual closure that is not part of any existing theory about levels of complexity, this book discusses a fresh take on well-known scientific concepts. Concepts are considered important, because they are the tools for scientific reasoning. The idea is that quality tools, the tools with as little as possible conceptual friction, make the work easy.

Further reading

Alvarez de Lorenzana, J.M. (1993). The constructive universe and the evolutionary systems framework. In: S.N. Salthe (ed.). Development and evolution. Complexity and change in biology. MIT Press, Cambridge, MA, USA. Appendix pp. 291-308.

Jagers op Akkerhuis, G.A.J.M. (2010). The operator hierarchy: a chain of closures linking matter, life and artificial intelligence. Dissertation, Radboud Universiteit Nijmegen, the Netherlands.

Jagers op Akkerhuis, G.A.J.M. (ed.) (2016). Evolution and transitions in complexity. The science of hierarchical organization in nature. Springer, Cham, Switzerland.

Miller, J.G. (1978). Living systems. McGraw-Hill, New York, NY, USA.

Simon, H.A. (1962). The architecture of complexity. Proceedings of the American Society for the Philosophy of Science 106: 467-428.

Szathmáry, E., Maynard Smith, J. (1995). The major evolutionary transitions. Nature 374: 227-232.

Waddington, C.H. (1969). Paradigm for an evolutionary process. In: C.H. Waddington (ed.). Towards a theoretical biology, Vol. 2: Sketches. Edinburgh University Press, Edinburgh, UK.

2. Selfish genes, powerful organisms

Genes or organisms. Genes are regions of the DNA that – alone or in combination – code for the phenotypic properties of an organism. In his book 'The Selfish Gene', Richard Dawkins places the gene center stage in questions about inheritance and evolution. But what should be viewed as the fundamental unit of selection in biology? The gene or the organism?

The selfish gene. Through their effects on the construction and physiology of the organism they reside in, genes can be viewed as 'steering' their organism in the same way as the driver of a car steers his 'vehicle'. The viewpoint that genes 'use' the organism for their own survival and replication, leads to Dawkins' anthropomorphism of the 'selfish' gene. Dawkins promotes such a gene-centered view of evolution between others to explain behaviors that seem to go against the goals of the organism as an individual. For example, adult salmon seem to act against their own survival instincts when they swim to the upper reaches of a river to spawn and die.

The viewpoint of selfish genes suggests that genes are the units of selection. Dawkins is of course aware that genes and organisms can both be viewed as units of selection, but he is ambiguous about whether one or the other should be preferred. Sometimes he writes that: '*The fundamental unit of selection, and therefore of self-interest, is not the species, nor the group, nor even strictly the individual. It is the gene, the unit of heredity.*' But on another occasion, he writes: '*On any sensible view of the matter Darwinian selection does not work on genes directly.*' In this SCIENCEBITE I evaluate whether ambiguity on this point is necessary. Maybe one can arrive at a conclusion suggesting that either the gene or the organism is the unit of selection?

Who is acting? A decade before Dawkins published The Selfish Gene, Conrad Hal Waddington stated that evolution depends on two things: a stable memory store, and a container-like environment, which Waddington named an 'operator' (Figure 2.1). The operator-environment maintains the genes/DNA and surrounds these with an interface in the form of an

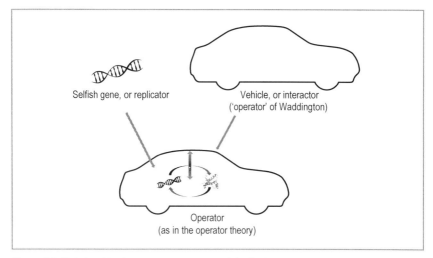

Figure 2.1. Relationships between the concepts of (selfish) gene, replicator, vehicle, interactor, and operator.

organism. The organism as a whole allows the genes to be indirectly active in the environment.

In contrast to Dawkins, Waddington emphasizes the organism-viewpoint, because he views the genes as dull memory stores, and the organism as the agent that is active in the world. As active agents, organisms can be viewed as profiting or suffering from interactions with the environment, including other organisms. Whenever an organism lives to reproduce, or dies before reproduction, it is always the organism that pays at the cash-desk of nature. In Waddington's view, it is the organism that is the unit of selection, while a gene is a unit of heredity.

An organism can function autonomously. In an environment where it finds the proper resources an organism can function autonomously. Genes can't do that. A gene is really just a segment of a DNA (or RNA) molecule. A gene cannot act autonomously. To be of any use as 'heritable information' a gene needs a cell that takes care of the maintenance of the gene's coding, of the deciphering of this code, and of its activation during the existence of the cell. For this reason, it hardly makes sense (in nature) to consider the evolution of genes without their hosts, the organisms. Of course, one can view the DNA/RNA of a viroid after it has dissolved in a cell, as having released a kind of 'naked gene', representing a molecular replicator. However, a naked virus gene cannot replicate autonomously. Without replicase enzymes in a test tube, without a cell, or without many cells of a multicellular organism, the viral genes cannot replicate.

The terms replicator and operator later inspired David Hull to link the gene as a replicator to an 'interactor', formerly called the 'vehicle' (Figure 2.1).

Broadening Waddington's operator concept. In the work of Waddington, an operator is a thing that surrounds the genes, and that allows the genes to, in an indirect way, be active in the world. While the operator theory was inspired by Waddington's use of the term operator, it uses the term operator in a broader sense than Waddington, for two reasons. Firstly, the term operator in the operator theory aims at addressing entities that form a physical and functional whole. For this reason the focus is on the entire organism. Secondly, in the operator theory I aimed at using a the term that could be applied equally well to organisms and to non-biological entities, such as molecules, atoms and hadrons, of which protons and neutrons are examples. As all such entities can be viewed as operating autonomously in their environment, the term operator seemed a logical choice.

The operators as defined by the operator theory also offer a means for distinguishing between things that are organisms and things that are not organisms. The viewpoint of the operator theory on what is an organism overlaps for a large part with the conventional views in biology. For example, a mouse, an elephant, and an earthworm are all organisms. In addition, the operator theory allows for more precision. For example, by focusing on dual closure, the operator theory excludes as organisms all kinds of organization that lack dual closure, for example lichens and the slug of a slime mold. These are no longer viewed as organisms but as compound objects that consist of several interacting organisms.

Every organism has parts, including the genes (the 'replicators' of Dawkins), and the surrounding material (the 'operator' of Waddington, or the 'vehicle' of Hull). The reason the operator theory does not split these parts is that, because of dual closure, the parts interact to form one operator/

organism, representing a single, functionally and structurally united physical unit.

The operator theory defines unity. Unity in the operator theory depends on three criteria. To explain these criteria I will use the bacterial cell as an example.

- First criterion: a limiting interface, which in the case of a bacterium takes the form of the membrane. The membrane as a structure is selectively closed for the passage of various kinds of molecules into and out of the cell. There is 'structural closure'.
- Second criterion: a cell has a self-maintaining set of autocatalytic molecules. Self-maintenance implies that all the catalytic reactions in the set mutually support each other. There is 'functional closure'.
- Third criterion: the membrane and autocatalytic process are mutually dependent. The mutual dependence implies that the autocatalytic set of a cell produces the material for the membrane, while the membrane mediates the flows of 'food' into the cell, and 'waste' out of the cell. Importantly, the membrane keeps the autocatalytic molecules together.

If the criterion of obligate mutual dependency is fulfilled, the system is said to have 'dual closure'. Dual closure of the cell implies the presence of (set-wise) autocatalysis and a membrane, and their mutual dependence. Dual closure leads to a special kind of unity, that both involves functional and structural aspects. In this way dual closure helps, for example, to specify what is 'auto' about autopoiesis.

General use of dual closure. Dual closure applies to cells and organisms, but it also applies to specific units that are less complex than organisms, such as molecules and atoms. As was explained in SCIENCEBITE 1, every unit that is defined by dual closure is called an operator. With quarks acting as the fundamental building blocks, the range of operators currently known include abiotic entities such as hadrons, atoms and molecules, and biotic entities such as cells, endosymbiont cells (the cells with an endosymbiont;

the 'eukaryotes'), multicellular organisms (made of cells, or of endosymbiont cells), and neural network organisms.

Even though essentialism has a bad name in philosophy, dual closure offers a minimal and 'essential' set of criteria for defining any operator. The 'kind' of the dual closure of any operator depends on the number of dual closures that preceded its construction. Such a number also indicates the level of complexity.

Interdependence. From the perspective of dual closure, a gene cannot be viewed as driving the organism's body as its vehicle. Instead, the genes and the body of the organism are interdependent parts of one and the same operator. Such interdependence makes it hard to selectively point at genes as units of selection. Basically, genes are important units of heredity. But genes are not the only structures that can be inherited. The cell's membrane and the cell plasma are also inherited, as are obligatory endosymbionts such as mitochondria or chloroplasts.

Viewing genes as the units of selection only represents a half-truth, because the survival or death of an organism does not depend primarily on its genes, but on the entire organism when it interacts with the environment. It is the organism that lives and can die. Because of this, one can conclude that the organism is the primary unit of selection in biology.

Units of selection. While the perspective of dual closure points at the organism as the unit of selection, a gene-centered viewpoint can still be relevant if the aim is to focus on genetic aspects of evolutionary phenomena, or when carrying out calculations on changing gene frequencies in populations. To avoid ambiguity, however, I suggest viewing the genes as heritable structures inside an organism, and acknowledging the organism as the unit that pays at the cash-desk of nature.

Additional levels. Compared to the genes of a bacterium, a protozoa has additional heritable structures in the form of mitochondria and/or chloroplasts. Both mitochondria and chloroplasts can reproduce as cells and harbor their own DNA. Accordingly, protozoa and other unicellular eukaryotes have two levels at which heritable structures play a role. However, because the mitochondria or chloroplasts are obligatorily bound to their host, and because the host pays at the cash-desk of nature, it is the host cell that is the unit of selection.

In eukaryotic multicellular organisms, e.g. plants and animals, the multicellular body adds a third 'level' of organization. As long as the cells depend obligatorily on their multicellular organization, and are connected through plasma channels, one can again say that the multicellular organism pays at the cash-desk of nature. For this reason it is the entire multicellular organism that represents the unit of selection.

Further reading

Dawkins, R. (1976). The selfish gene. Oxford University Press, Oxford, UK.

Dawkins, R. (2006). Chapter 1: Why are people? In: The selfish gene (30th Anniversary Ed.). Oxford University Press, Oxford, UK.

Hull, D.L. (1988). Interactors versus vehicles. In: H.C. Plotkin (ed.) The role of behavior in evolution. MIT press, Cambridge, MA, USA.

Jagers op Akkerhuis, G.A.J.M. (ed.) (2016). Evolution and transitions in complexity. The science of hierarchical organization in nature. Springer, Cham, Switzerland.

Plotkin, H. (1994). Darwin machines and the nature of knowledge. Harvard University Press, Cambridge, MA, USA.

Waddington, C.H. (1969). Paradigm for an evolutionary process. In: C.H. Waddington (ed.). Towards a theoretical biology, Vol 2: Sketches. Edinburgh University Press, Edinburgh, UK.

3. Synthetic biologists construct life

Synthetic biology, a modern branch of science. Topics studied in synthetic biology include: designing biosensors, manipulating and transforming cells for the production of new materials, developing synthetic cell components, and constructing synthetic life. This SCIENCEBITE focuses on constructing life.

Constructing is the work of engineers. The involvement of an engineer in a construction process starts with the desire for a product, e.g. a 'bridge'. Next, the properties of the bridge must be defined. A bridge can be described as: 'a construction connecting two parts of a road that are separated by water, a canyon, a road, etc'. Further specifications could be that the bridge must be made of bamboo, that the span of the bridge is two meters, and that the bridge must be strong enough to carry the weight of five people. With this information an engineer can explore different designs and decide which design can serve as the blueprint for the actual construction process.

What is life? The example of the bridge is my analogy for the aim of synthetic biologists to construct 'life'. In principle, the engineering of life seems a straightforward undertaking that involves the desire for a product, the identification of properties (a definition), specification of the details, exploration of designs and construction activities. Interestingly the situation is not as simple as this, because biologists have problems arriving at a conceptualization of the term 'life', despite the fact that life holds a central position in the naming of biology as a discipline. Life can be viewed, for this reason, as a challenging bio-concept in need of a definition.

Obstacles for defining life. Interestingly, there is no shortage of scientific definitions of life. In fact, Radu Popa, as well as Mark Bedau and Carol Cleland demonstrate that more than 100 different definitions are available. What is required for engineering, however, is a definition that can act as a foundation for constructing. As has been indicated by Serhiy Tsokolov, several obstacles stand in the way of arriving at such a definition:

1. Attempts to define life are frequently based on undefined terms. For example, most definitions of life make use of the organism concept. The organism concept, however, has no consensus definition.

2. The concept of a 'definition' is frequently confounded with a 'list of properties'. Normally a definition should distinguish all the relevant things from all the irrelevant things. The problem with lists of properties is that it is not always clear whether *all* criteria listed should be used, or only *some* of them. Neither is it guaranteed that if one uses all the criteria listed, this leads to the inclusion of all relevant cases and the exclusion of all irrelevant cases.

3. Frequently, a choice is made for using a minimal living system (the cell) as a basis for a general definition. Yet, a complex multicellular organism like a human is characterized by other, more, and more complex properties than are present in a bacterium. For this reason it is not likely that any definition of life that corresponds to a bacterial cell will automatically apply to more complex 'life forms'.

I like to add a fourth point to the listing of Tsokolov, namely the relevancy of focusing on what *kind* of term 'life' is. In a sentence like: 'The ocean is teeming with life', life is used as a metaphor. In other cases life refers to a classification, or to a property. Classes and properties have no physical representations. It makes little sense to tell someone: 'give me the life', or 'you can now release the life from the box'. The thing a person can release is an organism, such as a rabbit. The rabbit is an organized physical unity, called an organism. The rabbit 'is' not life, but instead 'has' the property of life.

Life: a general term. Does the above help engineers to better understand life and identify criteria for constructing life? As the examples of the rabbit demonstrates, life is general term. General terms cannot be constructed. What can be learned from this, is that engineers need a definition of a physical thing they can construct. This physical thing is the organism.

Constructing an organism. An organism is a physical thing. For this reason defining the 'organism' may seem easy. Yet, the situation is problematic. There is confusion. For example, at the time of writing this book, Wikipedia states that an organism is an *'individual entity that exhibits the properties of life'*. Meanwhile Wikipedia defines life as: *'...a characteristic that distinguishes physical entities that do have biological processes,...from those that do not,...'*. One can rapidly deduce that biological processes are a unique property of organisms, and that for this reason Wikipedia's definition is circular.

Wikipedia also cites Daniel Koshland's 'Seven pillars of life': *'All types of organisms are capable of reproduction, growth and development, maintenance, and some degree of response to stimuli.'* If one aims at a definition without exceptions, a listing of criteria like this is tricky. It is easy to find an exception demonstrating that the criteria don't *always* apply. For example, a frozen bacterium, and a desiccated seed are not capable of reproduction, are not growing, are not developing, are not involved in maintenance and are not capable of responding to stimuli. According to the criteria they can't be 'life'. And how about organisms that can't reproduce, such as a mule, a sterilized cat, or a grandmother?

Because the criteria of the seven pillars of life do not always apply, they can be viewed as what Ludwig Wittgenstein called 'family resemblance criteria'. A person can have the nose of his/her mother, the eyes of his/her father, and so forth. To be recognized as part of the family a person doesn't need to have all family resemblance criteria at the same time. Current listings of criteria for life seem to be used as family resemblances. If a member of the family of the organisms must not comply with all criteria, this helps to deal with a mule, a frozen bacterium, or a grandmother. However, if one relies on family resemblance one risks including as organisms an unknown number of inappropriate cases, such as a flame, because it has metabolism, or a virus and a computer virus, because they can reproduce.

The latter problems demonstrate that it pays to search for a more conclusive list of criteria for defining the organism and/or life. Maybe it is possible to follow a new and different approach?

Organism blueprint. The above indicates the need for an approach that describes an organism regardless of whether it is active or not, or is capable of reproducing or not. It is relevant in this context, that the approach of the operator theory does not depend on activity. The operator viewpoint makes use of a recursive definition in which low level operators form the basis for the construction of higher-level operators, in a long sequence starting with the quarks.

In the hierarchy of the operators, every next higher-level operator is produced through special interactions between lower level operators. These special interactions must meet the criterion of 'dual closure'. Dual closure implies the presence of a structural closure and a functional closure, in mutual dependency (see also SCIENCEBITE 1). Starting with quarks, a series of successive dual closures in nature has produced hadrons, atoms, molecules, cells, endosymbiont cells (cells with an endosymbiont), multicellular organisms (of cells, and of endosymbiont cells), and neural network organisms. If one uses this hierarchy it becomes relatively easy to define the organism concept as follows: *any operator of a kind that is at least as complex as the cell.*

Defining life. A new approach was suggested above for defining the organism. And the definition of the organism can now be used for defining life in two ways. The first option is to focus on a property named 'organismic life', or O-life (Figure 3.1). O-life refers to the *highest-level* dual closure in an organism. The term organism was defined above. The focus on the highest-

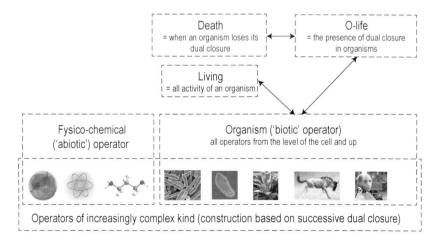

Figure 3.1. An overview of the kinds of physical entities and theoretical constructs that play a role in the definitions of 'organism', 'O-life', 'living', and 'death'. Dashed lines indicate theoretical classes.

level dual closure is important. For example, when dementia takes away the dual closure of a person's brain, he/she no longer functions as a neural network organism (a 'memon'), even though with proper care his/her body can survive as a 'multicellular' operator, as a 'plant'. Without care, the cells of his/her body would die, leaving behind a remainder called a corpse.

The second option is to focus on a system and define 'systemic life' (or 'S-life'). S-life refers to a system in which one finds at least one organism that interacts with its environment. Reproduction and evolution are optional in an S-life system.

Both the concepts of O-life and S-life refer to classes, not to physical things. The physical basis for both life-concepts is the organism.

The task at hand. The above reasoning can help solve the riddle of how to engineer 'life'. First of all, I would suggest that what engineers aim at constructing is an organism, instead of life. Saying that one will 'construct life' must be interpreted as a metaphor. Second, the blueprint that can describe any and all organisms is the presence of dual closure. Third, the organisms with the least complicated dual closure are bacteria/archaea.

In SCIENCEBITE 1 I explained that dual closure of a bacterium implies three things: (1) a membrane; (2) an autocatalytic set; and (3) obligatory mutual dependency of the autocatalytic set and the membrane. If synthetic biologists manage to construct a thing that complies with these three basic criteria, they have constructed a cell, which has the property of life, and which, when active, is a living being.

With a minimal blueprint at hand, the next step is to identify chemical self-organization processes that can bootstrap the formation of a simple cell, possibly in interaction with a scaffolding of porous rocky material. While this is undoubtedly a marked challenge that requires advanced engineering, a lean approach would imply that simplicity may be a better guidance than complexity.

Both the *de novo* formation of a simple cell from an abiotic chemical environment, and the construction of an engineered cell from high level components lead in different ways to the goal of engineering a cell, as the simplest kind of organism. Meanwhile, the construction of a bacterial cell from elements that are extracted from a complicated, late evolutionary bacterium, offers more insight, but does not answer the question of how the first cell(s) came about. The question of 'abiogenesis' is not answered.

Further reading

Bedau, M.A., Cleland, C.E. (eds.) (2010). The nature of life: classical and contemporary perspectives from philosophy and science. Cambridge University Press, Cambridge, UK.

Jagers op Akkerhuis, G.A.J.M. (2010). Towards a hierarchical definition of life, the organism, and death. Foundations of Science 15: 245-262.

Jagers op Akkerhuis, G.A.J.M. (2010). Explaining the origin of life is not enough for a definition of life. Foundations of Science 16: 327-329.

Jagers op Akkerhuis, G.A.J.M. (2012). The Role of Logic and Insight in the Search for a Definition of Life. Journal of Biomolecular Structure and Dynamics 29: 619-620.

Jagers op Akkerhuis, G.A.J.M. (2016) Evolution and transitions in complexity. The science of hierarchical organization in nature. Springer, Cham, Switzerland.

Koshland, D.E. Jr (2002). The seven pillars of life. Science 295: 2215-2216.

Popa, R. (2004). Between necessity and probability: searching for the definition and the origin of life. Advances in Astrobiology and Biogeophysics. Springer, Cham, Switzerland.

Tsokolov, S.A. (2009). Why is the definition of life so elusive? Epistemological Considerations Astrobiology 9: 401-413.

Wittgenstein, L. (2001). Philosophical investigations. Blackwell Publishing, Hoboken, NJ, USA.

4. Strong versus operational definitions of life

Defining life. Recently Leonardo Bich and Sara Green published the following study: *'Is defining life pointless? Operational definitions at the frontiers of biology'.* The paper offers an overview of the current debate about whether or not it is possible and/or useful to define life. What I find attractive in this study is that it offers a comparison of the characteristics that are typical for so-called 'strong' definitions and 'operational' definitions, which are summarized in the following lines (after Bich and Green, 2018, Table 1).

Strong definitions. Strong definitions of life have the following properties: (1) they make strong ontological claims. With these claims they aim at demarcating life in a way that can be viewed as carving out natural kinds; (2) they work towards complete definitions, that offer necessary and sufficient conditions; (3) the definitions of strong approaches are generally fixed in the sense that they propose static categories; and (4) strong definitions aim at consensus and generalization for all life of the characteristics that are addressed.

Operational definitions. Operational definitions of life differ from the above four points in a number of ways: (1) the claims that are made are instrumental (instead of ontological) and the descriptions are viewed more as theoretical and epistemic tools that assist in the guidance of debate and/or experiments; (2) the definitions are open-ended (no final definition is aimed at) and it is viewed as more important to offer necessary conditions that may be provisional (without having to be sufficient); (3) operational definitions are not fixed and may change over time; and (4) the practical utility is the main goal of operational definitions, which for this reason remain a constant target for debate, challenge and revision. Different approaches can lead to different definitions at the same time. The simultaneous existence of multiple points of view demands a pluralistic environment for debate.

Valuing the qualities of both perspectives. In research it happens many times that a specific subject can be looked at from contrasting angles. For example, the same entity can be viewed simultaneously as a system, and as an object (for which reason I introduced the term 'sysob' in Gerard Jagers op Akkerhuis, 2018). Likewise one can focus on genes or on organisms as the units of selection (see SCIENCEBITE 1). Or one can contrast Darwin's view of evolution with that of the Modern Synthesis or of the Extended Synthesis.

In situations like these, the debate can easily become heated and the parties involved may split into opposing camps. Normally such camps reach reconciliation at some point. This process can also be described by the triad: thesis, antithesis, synthesis, which can be found in the works of philosophers like Immanuel Kant and Georg Wilhelm Friedrich Hegel.

The above summary after Bich and Green suggests a divergence between the strong viewpoint and the operational viewpoint. With this essay I aim to contribute to a reconciliation; a synthesis viewpoint that combines valuable aspects of both extremes. The reward of such a reconciliation is that the result allows for approaches to defining life that combine strengths and operational value.

Interpretative variability. People often fear that definitions will limit their options. This does not have to be the case if one focuses on the 'interpretative variability' of definitions. Such a focus allows one to identify the criteria of different viewpoints and study whether or not they can be combined.

As a very open approach to describing what is a definition, I view a definition as a 'chunk' that is cut out of an infinite possibility-space. This chunk can be closed, like a ball. Or it can be open, like an endlessly large funnel. Some chunks focus on things that exist in the here and now, like 'this stone'. Other chunks focus on events, such as 'walking'. Definitions can be specific about the 'kind' of thing one talks about, or not. For example, the color 'red' is of the kind 'property'. The property red refers to a class of colors recognized as red, and a corresponding class of wavelength that an object may emit. Red is never a thing. One cannot say: 'Give me a red'. However, one *can* say: 'Give me a red ball'.

For me, the utility of discussing definitions is that such discussions help create a shared understanding of a specific subsection of possibility-space. The aim is to create clarity about the 'thing' one has in mind, and its 'kind'. Not all definitions serve the same purpose, and not all definitions are equally good. If one aims to compare the quality of definitions one can use several principles. An example of a basic principle is Ockham's razor, which is currently interpreted as: one must look for the simplest yet sufficient approach. Another principle is generality, which means that one must look for definitions that are valid for any and all relevant cases.

What is life? Before discussing strong or operational definitions of life, one has to agree on the 'kind' of the concept. Above I indicated that one cannot say: 'Give me a red'. A physical thing called a 'red' does not exist. Red is a property. By analogy one cannot say: 'give me a life'. This is because the concept of life does not refer to a physical thing.

This insight has consequences for studies in which laboratory experiments are carried out to create life, or efforts are directed at the modeling of life, in which cases the concept of life is used in a metaphorical way. What is physically created or modeled is not 'life' but an organism, or a constellation of interacting organisms, or some physical part of an organism. The upshot of this reasoning is that the concept of life can be viewed as referring to a generic organizational property all organisms have. To define life as a property of organisms, one must know what an organism 'is'. So, what defines an organism?

What defines an organism? Biologists have difficulty with defining the organism. In analogy with life, one can summarize strong and operational definitions of the organism. As organism are physical things, one may expect that in the end strong and operational approaches must overlap. After all, a lack of overlap would indicate inconsistent logic.

As I see it, the major difficulty of defining the organism lies in the identification of criteria that are valid for any and all organisms. Two obstacles are in the way of this goal: (1) different kinds of organisms may require different criteria; and (2) as long as there is no generally accepted definition of the organism concept, it is not possible to start with a selection of things that are organisms and from this basis deduce a generally valid definition. In the absence of a general definition, it is possible that several things that are currently called organisms may lose this status in due time.

Both the understanding of which kinds of organization can be called an organism, and the grouping of things that are organisms, could profit from an external framework for differentiating between things in the world that are organisms and things that are not.

While other frames of reference may be suggested, this book makes use of the operator hierarchy. The ranking of this hierarchy runs from quarks to neural network organisms and can be used with little difficulty to define as organisms: *every operator of a kind that is at least as complex as the cell*. The resulting organism concept refers to a specific subset of high-complexity operators, while excluding any and all other entities.

Practical examples. How can the above definition be put into practice? For example, every single bacterium/archaeon classifies as an organism because it is an operator of the kind cell. And because they are all operators of the kind 'endosymbiont cell' (biologists refer to this kind of operator as 'eukaryote cell'), all protozoa classify as organisms too. In the same way, higher-level organisms can be identified, such as plants and animals with brains.

The above organism definition acknowledges that a life-cycle may involve different kinds of organization that must not all be organisms. For example, the human zygote is a single cell, and an organism. When it divides, a group forms of 2, 4 and 8 clonal cells. This group of cells is not one organism, but a group of organisms. It is only when the plasmas of the cells become connected, and the cells start mutually influencing each other's physiology through these plasma connections, that dual closure can be recognized, and the cells become a multicellular operator, and a single countable organism, in the context of the operator theory.

Reconciling strong and operational definitions of 'life'. Now that the organism concept has been defined, the concept can serve as a foundation for two specific definitions of life: an organismic and a systemic definition. Here I focus on the organismic definition, which I refer to as O-life: *O-life is a concept indicating the presence of dual closure in any organism*. As I explained in SCIENCEBITE 1, dual closure, when defined in its smallest form, combines three criteria: (1) functional closure; (2) structural closure; and (3) Mutual dependence of these two closures. This definition is more specific than one that suggests that life refers to a state of matter, because the presence of dual closure details *which* state of matter is relevant.

The loss of the property O-life implies that dual closure is lost. In most cases the loss of dual closure is synonymous with the death of the organism. Death means that the thing that once was an organism, changes into a corpse. When I use 'in most cases' in the above sentence, this is to leave room for a step downward on the operator ladder, for example from a multicellular plant, to single plant cells dwelling in a culture broth. In such cases splitting up the bonds between the plant cells does break the dual closure but does not lead to a corpse. Instead what result are cells that remain fully functional, but now at the unicellular level.

As long as dual closure is intact, both a frozen organism and a desiccated organism still have the property of O-life. One may now remark that classical definitions assume that life is associated with reproduction, metabolism, reactivity, etc. However, when using such criteria in the case of a frozen or desiccated organism, one would have to conclude that the frozen/desiccated organism does not represent life, because it lacks such properties. Yet, a desiccated seed, and a frozen tree frog still have the organizational properties of O-life. After thawing/wetting they can become active again.

Strengths of O-life. O-life is a definition that is both minimal, strong, and operational. It is minimal because it only uses three criteria. It is strong because only organisms comply with this definition. It is operational because dual closure is linked to physical phenomena that can be studied in all organisms, from the autocatalytic chemistry and membrane of a bacterium, to sensors and neural functioning in complicated multicellular organisms like a human.

Because of these properties it seems to me that the use of O-life can reconcile the camps of scientists who are now split along the lines of strong or operational definitions.

Further reading

Bich, L., Green, S. (2018). Is defining life pointless? Operational definitions at the frontiers of biology. Synthese 195: 3919-3946.

Jagers op Akkerhuis, G.A.J.M. (2010). Towards a hierarchical definition of life, the organism and death. Foundations of Science 15: 245-262.

Jagers op Akkerhuis, G.A.J.M. (2016). Learning from water: two complementary definitions of the concept of life. In: G.A.J.M. Jagers op Akkerhuis (ed.). Evolution and transitions in complexity. The science of hierarchical organization in nature. Springer, Cham, Switzerland.

Jagers op Akkerhuis, G.A.J.M. (2018). Een realistische, trans-disciplinaire ontologie voor de empirische basis van de wetenschap: bijdragen van de operator theorie. [A transdisciplinary realist ontology for the empirical basis in science: contributions of the operator theory]. https://doi.org/10.13140/RG.2.2.12851.12321

5. Is 'the origin' about species?

Darwin's Origin. The complete title of Charles Darwin's masterpiece (1859) reads: 'On the origin of species by means of natural selection, or the preservation of favored races in the struggle for life.' When I use the word masterpiece I declare myself a great fan of Darwin. Darwin (1876) really surprised me when I discovered that he had considered using Pierre-Louis de Maupertuis's work on the least action principle. The idea of linking this fundamental principle to evolution is something that – as far as I know – few current evolutionists consider (possibly they think that the least action principle is a physical principle, but it is equally important for organisms). But let me return to the 'Origin' (Figure 5.1).

Figure 5.1. 'On the origin of species' (Darwin, 1859).

Evolution. Darwin writes the 'Origin' in a time when many people believe that God created groups of like organisms, in neat boxes called species. Darwin develops an alternative view in which a single universal ancestor over many generations of descendants diversifies into a broad range of organisms of different species. What Darwin aims at is a *naturalistic* explanation for all the many species. Darwin does not mention evolution in the 1859 version of the Origin and uses 'evolution' twice in the 1876 version. Darwin uses evolution in close association with his theory of *'descent with modification through variation and selection'*. Here I will take an interest in the kinds of entities these terms can refer to.

Is the 'Origin' about species? The phenomenon we now think of as evolution, is what Darwin refers to as the theory of *'descent with modification through variation and selection'*. By choosing these words Darwin leaves open the possibility that each term can represent either a verb or a state. For example, the term modification can refer to the process of modification, or to the result; a modification, a change. Such ambiguity leaves much freedom for applying these terms to different kinds of things, species and organisms alike.

When Darwin uses terms in the above way, a specific kind of logical twist, called a category mistake, lurks around the corner. This twist has to do with categories of things.

For example, an ecologist can say that all the species of the savannah come to drink from a local pool. What really happens is that the *organisms* of the species visit the pool for drinking. If one would say that a species can drink, this would be metaphorical use of language, because the category involved is a species, which is a 'grouping', a mental construct. A mental construct cannot do anything physically, such as drinking. Only a physical organism can drink water from a pool.

The relevance of categories. In my view, paying attention to categories is relevant for the study of evolution. For example, one can focus on an organism and the things it can do. An organism – as a physical being – is born, after which various aspects of its body may change during its life. Eventually the organism dies. Does an organism ever evolve? A check of Darwin's criteria teaches us that physically speaking, an organism is constructed by its parent. After being born, it can change its appearance, which may count as 'modification'. An organism cannot 'do' variation or selection as these are comparative assessments. The conclusion of this analysis is that an organism cannot evolve.

How about a species? Can a species evolve? A species is a group concept, and as such consists only in the mind. As we saw above, a group concept cannot act physically. Assuming that evolution refers to something physical, some kind of process in the real world, and recognizing that a species is a mental grouping instead of a physical thing, one must conclude that a species cannot evolve physically.

A focus on categories thus suggests that neither organisms nor species (and for the same reasons populations) can comply with Darwin's criteria for evolution. Maybe there is another perspective that can resolve this evolutionary puzzle?

A focus on species. When explaining evolution, what does Darwin actually define? To examine this, I suggest focusing on Darwin's explanation of evolution as a branching tree (Darwin, 1859, figure in between pp. 116-117): '*Let (A) be a common, widely-diffused, and varying species, belonging to a genus large in its own country. The little fan of diverging dotted lines of unequal lengths proceeding from (A), may represent its varying offspring.*' (the underlining is from the current author). Does Darwin say that species (A) produces one or more offspring-species? Or does Darwin imagine a group of organisms – metaphorically indicated as species (A) – of which each individual organism may reproduce, and produce physically varying offspring organisms? From an analytical perspective, Darwin may run into trouble in both cases.

The assumption that species (A) produces varying offspring-species implies a description in terms of logical groupings ('species' and 'offspring-species'). Logical groupings, however, have no physical agency, and cannot offer a materialistic explanation. Species cannot produce offspring.

Alternatively, one can assume that Darwin uses the species concept metaphorically, suggesting that physical parents produce physical offspring. The question that arises now is: when does the production of offspring by organisms in a parental population imply Darwinian evolution?

Are species things? It is relevant at this point to indicate that Darwin is aware of the mental nature of species. Darwin (1859, p. 119) states that: '*In our diagram the line of succession is broken at regular intervals by small numbered letters marking the successive forms which have become sufficiently distinct to be recorded as varieties. But these breaks are <u>imaginary</u>, and might have been inserted anywhere, after intervals long enough to allow the accumulation of a considerable amount of divergent variation.*' (the underlining is from the current author). The concept of a variety thus has theoretical limits that depend on '*a considerable amount of divergent variation*'.

With respect to species – which one can view as the largest possible varieties – the loss of the capacity of individuals of species (A) to mate and produce fertile offspring with individuals of species (B), is nowadays generally accepted as 'sufficient variation'. It is rarely possible to test these criteria for every individual of a species, even if one would limit the demands to mating with a single partner. People thus generally *assume by induction* that if some individuals can mate and produce fertile offspring, this holds for all similar-looking male/female organisms in the selection.

A focus on organisms. Whenever Darwin uses the word species, or variety, while explaining evolution, it is necessary – from a philosophical point of view – to check the use of categories. In this context it is interesting that the Origin contains a passage where Darwin explains the basis of evolution in terms of individuals only.

In Chapter XV, Recapitulation and Conclusion, Darwin (1876) writes: '*Nothing at first can appear more difficult to believe than that the more complex organs and instincts have been perfected,...by the accumulation of innumerable slight variations, each good for the <u>individual possessor</u>. Nevertheless, this difficulty...cannot be considered real if we admit the following propositions, namely, that all parts of the organisation and instincts offer, at least, <u>individual differences</u> – that there is a struggle for existence leading to the preservation of profitable deviations of structure or instinct – and, lastly, that gradations in the state of perfection of each organ may have existed, each good of its kind.*' (the underlining is from the current author).

This time Darwin focuses strictly on organisms, their properties, and variation in their survival or mortality, as the *causes* of evolution. When stated this way, organisms do not 'do' evolution but perform activities that have evolution as a result.

Darwinian evolution? We have observed that Darwinian evolution – defined as descent with modification through variation and selection – is something neither an organism, nor a species can do. A species cannot do it, because it is a mental grouping. And an organism cannot do it, because it can only survive, reproduce and/or die, but cannot show variation or selection.

Darwinian evolution is frequently called a process, and it seems that Darwin writes about evolution as a process too, but the exact category of 'Darwinian evolution' still remains an enigma after this SCIENCEBITE. In the next SCIENCEBITE I will explore whether or not this enigma can be resolved.

Further reading

Darwin, C.R. (1859). On the origin of species by means of natural selection, or the preservation of favoured races in the struggle for life (1st Ed.). John Murray, London, UK.

Darwin, C.R. (1876). The origin of species by means of natural selection, or the preservation of favoured races in the struggle for life (6th Ed., with additions and corrections). John Murray, London, UK.

De Maupertuis, P.L. (1744). Accord de différentes loix de la nature qui avoient jusqu'ici paru incompatibles. [Accord between different laws of nature that seemed incompatible]. Bruyset, Lyon, France.

6. Evolution: process or pattern?

Evolution: a process? Most people associate evolution with change and change with process. The process-view is currently the standard in evolutionary thinking. But there is a problem. Several criteria for Darwinian evolution, such as variation and selection, refer to concepts that are not processes, but that can be viewed as evaluations or assessments. Can evolution still be called a process if some of its criteria do not count as a process? Answering this question is relevant for our understanding of the concept of Darwinian evolution, and for finding a precise and potentially more generally acceptable definition.

Evolution: an umbrella term. In everyday practice people currently use evolution as an umbrella term. An umbrella term links the same word to different interpretations. Evolution has quite a few interpretations. Here I distinguish between the following: Charles Darwin's interpretation, the Latin interpretation, calculations based on the modern synthesis, and the 'Price equation' of George Price. For each of these options, I analyze whether or not the kind of thing evolution refers to actually is a process.

Darwin's Origin. Darwin's 'On the Origin of Species' remains a hallmark in the thinking about evolution. Darwin writes the book at a time when many people believe in a God who created organisms of different 'kinds', each kind named a 'species'. Accordingly, species are static, eternal. Following their creation, the organisms of each species always produce offspring that look similar to the parent(s).

Darwin suggests a different explanation: a theory of evolution based on *'descent with modification through variation and selection'*. Starting with a common ancestor, reproduction and differential mortality cause a pedigree, the 'tree of life'. Current species reside at the tips of the branches of the tree of life. The following quotes make me think that Darwin views evolution as a process: *'The many species have been evolved...', '...species have been evolved by very small steps', 'Everyone who believes in slow and gradual evolution,....', or '... the steps by which these curious organs have been evolved,....'.*

Darwinian evolution. Could it be that our intuition of viewing evolution as a process misguides us, and that the conventional criteria of reproduction, variation and selection, keep us away from defining Darwinian evolution in a more precise way?

To look for an answer to the latter question, I start with an analysis of Darwinian evolution. Darwin uses evolution in the context of the search for an explanation of how species come about. In SCIENCEBITE 5 I already discussed the conceptual implications of viewing organisms or species as the units of evolution. Here I combine an organism-based perspective with the use of pedigrees. The goal is to develop a definition of Darwinian evolution that respects the logical kind of every criterion involved, notably reproduction, variation and selection.

Evolution criteria. To define what he writes about as evolution Darwin uses four criteria: descent, modification, variation, and selection. The use of these criteria has consequences. When observing a single bacterium, one cannot speak about Darwinian evolution, because none of Darwin's criteria applies. If the bacterium splits, it produces two daughter cells. Assuming both daughter cells are identical to the parent, the parent-offspring relationship classifies as descent, but between parent and offspring there is no modification of the bodily forms. The four criteria of Darwin are not met. It can also happen that one of the daughter cells is different. Now one can identify descent, modification of the parental form (because there is a difference between parent and one offspring), and variation between the two offspring. Still not all four criteria are met. Next, for recognizing selection, one needs criteria. Darwin (1859) defines selection as '...*preservation of favorable variations and the rejection of injurious variations...*'.

Is selection normative? By linking preservation to favorability, and rejection to injuriousness, Darwin adds a normative perspective (favorability/injuriousness) to empirical terminology (preservation/rejection). From a philosophical perspective, such links can lead to friction in the logic, because an organism can possess many properties that on average are 'favorable', and still die from a random cause, for example when being buried by an avalanche, or when being struck by lightning. Whether or not a property is favorable or injurious can in the individual case only be established the moment the organism dies or reproduces.

To create an empirical framework that is free of normative connotations, I think that Darwin could simply have suggested that selection refers to the difference between the survival (Darwin speaks about preservation) of some variations, versus the death/perishing (Darwin speaks about rejection) of other variations. In this way one uses only empirical and individual-based terms. Now it is safe to interpret survival/preservation as an event in which an organism lives until it produces offspring, whilst death/rejection refers to death before the organism reproduced.

Even if one uses a framework in which only actual events play a role, such as death and reproduction, one can always reconstruct the link with normative statements about the 'favorability' of a specific variation. To establish this link, one can calculate the average over many empirically observed patterns of descent and use the result to judge *a posteriori* whether or not specific variations were generally 'favorable' because they allowed the average individual to perform well, to be 'fit', and to survive and reproduce in a specific environment.

A Darwinian pedigree. Let's return to the bacterial example at the moment that the parent cell just produced two daughter cells, while these daughter cells differ in properties. And let's assume that each of these daughter cells produces two daughter cells (Figure 6.1). In this case both the first-generation daughter cells (B_1 and B_2) are 'preserved' because each of them produces offspring before dying. No 'rejection' occurs. As there is no rejection, Darwin's criterion of selection is not met.

Things change if only one (B_2) of the two different daughter cells produces daughter cells, while the other cell (B_1) dies before splitting (Figure 6.2). Now one can observe selection based on the preservation of one variation and the rejection of another variation. The pedigree meets all the criteria for Darwinian evolution. From this basis more complex pedigrees can be imagined, including pedigrees for sexually reproducing organisms.

Darwinian evolution. To better understand the concept of Darwinian evolution it helps to determine what kind of concept one is talking about. To

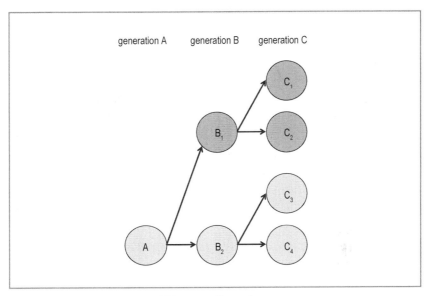

Figure 6.1. A bacterial pedigree in which every cell reproduces.

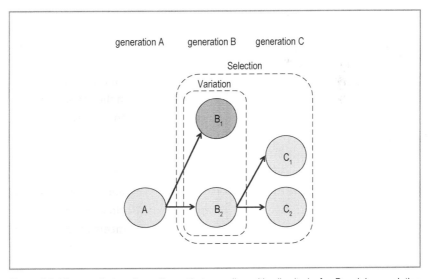

Figure 6.2. The smallest-scale pedigree that complies with all criteria for Darwinian evolution (assumptions: reproduction is asexual and takes place through division).

find out about this, I take a close look at variation and selection in the bacterial example of Figure 6.2. Any analysis of variation starts with a comparison of the properties of the two daughter cells, and continues with a test of whether or not these properties are different. The involvement of a 'difference' implies that one cannot decide about the presence or absence of variation without carrying out an assessment.

By analogy, also Darwin's criterion of selection asks for an assessment: it must be decided whether or not the daughter cells differ in two ways: 1. in their properties and 2. in achieving reproduction.

Summing up, Darwinian evolution involves three aspects: (1) the production of offspring: this is a process by any means; (2) variation between daughter cells: this involves an assessment of differences; and (3) selection: this involves an assessment of differences in properties and how these relate to mortality before reproduction. Of the latter three criteria only offspring production classifies as a process, while variation and selection classify as assessments.

According to the above, Darwinian evolution can be defined as the combination of one process and two assessments. When viewed this way, it becomes clear that Darwin's criteria in fact define a *pedigree with selection* (selection as a differential measure, not as a process). A pedigree like this can be described as a specifically patterned graph (graph = nodes connected by lines). Realizing this, I became aware that Darwinian evolution is not the process I always thought it was. I now think of Darwinian evolution as a special graph pattern, or simply as a pattern.

Latin interpretation and modern synthesis. Above I mentioned several homologies of evolution. Are these processes or patterns? A question like this requires that each case be investigated. The first, the Latin viewpoint, refers in a general way to 'unrolling' which implies a *process*. There is no demand for variation or selection. Any pedigree that results from offspring production, or more generally, any process in which things change appearance, can be viewed as a kind of unrolling.

The second, Darwinian evolution, classifies as a pattern. It is a pattern, because one needs a pattern for combining one process (reproduction) and two assessments (variation, selection).

The third case refers to calculations in the context of the modern synthesis, which link evolution to a change in gene frequencies in a population over generations. Now imagine that a parent bacterium produces two daughter cells, one having a mutation. The mutant produces 3 instead of 2 offspring. The mutation is heritable (Figure 6.3).

Assuming favorable conditions, the bacteria always manage to reproduce. Generation after generation the original genotype produces two offspring, while the mutation produces three offspring. The fraction of mutants rapidly increases. If one focuses on the change in gene frequencies, the conclusion would be: evolution.

However, as long as every bacterium manages to reproduce one cannot speak of the preservation of some versus the rejection of other variations. This means that selection in the Darwinian sense does not occur. Darwin would conclude: no Darwinian evolution. Apparently Darwin's criteria and those of the modern synthesis overlap only in part. Partial overlap implies

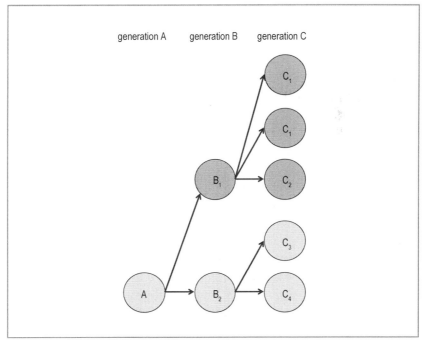

Figure 6.3. A bacterial pedigree in which a mutation produces three offspring instead of two.

that it is no longer logically consistent to assume that both approaches define the same thing.

Acknowledging the differences, I view the approach of the modern synthesis as a *measure* for (differential) genetic change when averaged for all individuals in a selected group. And I view Darwin's approach as aiming at the identification of necessary and sufficient criteria for *defining the concept* of Darwinian evolution.

Branching. Above I identified two distinct interpretations of 'selection':
- (A) Selection as an increase/decrease of the fraction of genes. This kind of selection mirrors 'fitness' in terms of the number of offspring.
- (B) Selection as a property-dependent difference between offspring in the realization of reproduction before death. Now selection is interpreted after Darwin.

In earlier work, I suggest referring to pedigrees that exhibit no mortality before reproduction, as 'branching'. When branching co-occurs with a change in the fractions of genotypes, as in the bacterial example with two or three offspring (Figure 6.3), I suggest addressing this situation as 'differential branching'. Branching doesn't comply with Darwin's criteria, because branching does not include cases of mortality before reproduction, and – for this reason – does not qualify as a pattern of selection. Of course, a branching pattern can always turn into a Darwinian pattern when in any generation mortality occurs in a differential way.

Price equation. The difference between Darwinian evolution and differential branching is relevant for understanding the role of the Price equation in biology. The Price equation quantifies the relative increase in the abundance of a genotype. The focus is on fitness, without the need to distinguish the two kinds of selection (A and B) discussed above. Darwin's criterion of selection is not used explicitly. The result is that Price offers a measure for genetic change in a population. Meanwhile, Darwin aims at offering necessary and sufficient criteria for Darwinian evolution. I conclude that Price and Darwin aim at achieving different goals: Price aims at a quantitative *measure* for differential changes in gene frequencies that can occur as the result of branching and/

or selection. Darwin aims at a *definition* of evolution in a form that includes selection.

Further reading

Brian Arthur, W. (2009). The nature of technology: what it is and how it evolves. The Free Press Simon & Schuster Inc., New York, NY, USA.

Darwin, C.R. (1859). On the origin of species by means of natural selection, or the preservation of favoured races in the struggle for life (1st Ed.). John Murray, London, UK.

Darwin, C.R. (1876). The origin of species by means of natural selection, or the preservation of favoured races in the struggle for life (6th Ed., with additions and corrections). John Murray, London, UK.

Jagers op Akkerhuis, G.A.J.M. (2016). Evolution and transitions in complexity. The science of hierarchical organization in nature. Springer, Cham, Switzerland.

Price, G.R. (1970). Selection and covariance. Nature 227: 520-521.

7. *Natura facit saltus*

Small changes. According to Charles Darwin's theory all present-day species are the product of the accumulation of small changes over many generations. As small changes scaffold larger changes, nature doesn't make leaps. Or does she?

Skyhooks. Daniel Dennett always emphasizes that there are no skyhooks in nature. The 'no skyhooks' statement implies that ever since a still unknown origin, all things are constructed from the bottom up. This also holds true when engineers design an airplane, and construct it from physical parts. The statement of Dennett is in perfect harmony with the classical statement '*Natura non facit saltus*' (nature doesn't make leaps). The absence of leaps also fits in perfectly well with the thinking of the mathematician Gottfried Wilhelm von Leibniz (1646-1716), one of the inventors of infinitesimal calculus. The 'no leaps' criterion is also part of Darwin's evolution theory.

Leaps. To gain an understanding of the concepts 'continuity' and 'leap' one can look at a yardstick. A yardstick is a continuous piece of metal or wood. A yardstick generally has small tick-marks and larger tick-marks, e.g. for millimeters, centimeters, meters. A tick-mark indicates a logical arrangement based on 'distance'. A yardstick thus combines two worlds: the continuous physical world of a piece of metal or wood, and the logical world of the tick-marks. In a similar way, many physically continuous processes can be divided logically into 'steps', 'leaps', or 'transitions'.

When creating a yardstick, it is relevant that the tick-marks are ranked according to some kind of order. However, order does not automatically suggest directionality. The fact that tick-marks on a yardstick indicate preceding or next steps, does not necessarily imply that one has to follow the yard-stick in one direction only. One can theoretically always go 'up' or 'down' a yardstick.

Embryonic development. For analyzing biological processes, biologists use tick-marks in many different ways. As an example one can analyze the development of a mammalian embryo, such as from a human or a pig, starting with a zygote. The zygote divides. The two daughter cells divide again, and again, until a group of eight adhering cells has formed. Normally around this time a special event takes place: the eight cells connect through plasma connections (called gap junctions), hereby producing an entity that consists of eight cells with a connected plasma. The relevancy of the plasma connections for the embryo can be demonstrated by adding a drug that blocks their formation. Now, the embryonic development proceeds in an abnormal way.

Plasma connections are not only typical for animals. They are also found in multicellular plants and algae.

Leaps during embryonic development. Figure 7.1 describes the early development of the mammalian embryo. The most detailed tick-marking focuses on generations of cells. The intermediate tick-marks focus on phenotypic stages: a single cell, a lump of cells (a pluricellular colony), and cells with connected plasma (a multicellular organism). The most abstract tick-marks focus on the kinds of organisms involved, e.g. unicellular or multicellular organisms. The more abstract a classification, the more details are left out, and the larger the 'jumps' that separate tick-marks. If one focuses on such large tick-marks it seems as if nature does make leaps. In general this is a delusion resulting from the use of abstract classes.

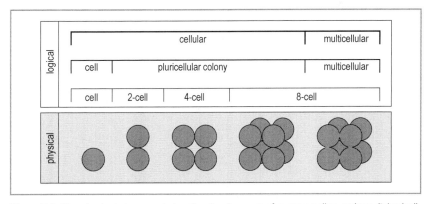

Figure 7.1. The physical changes during the development of a mammalian embryo ('physical', bottom) and a comparison of different methods of logical tick-marking ('logical', top).

'Leaps' in Darwin's tree of life. The classical picture of evolution has the appearance of a branching tree. Every branch represents a grouping called a 'species'. As a general rule, branches diverge. Exceptionally, however, two branches merge. Konstantin Mereschkowski and Boris Kozo-Polansky were biologists who started paying attention to the merging of branches, naming the process 'symbiogenesis'. An example of a merger is the incorporation of mitochondria in a host cell. This merger is responsible for the formation of the eukaryotes. A subsequent merger is the additional incorporation of chloroplasts during the formation of unicellular algae.

Not only does Darwin's tree include mergers between branches, mergers in one and the same branch also happen. For example, clonally produced cells can form plasma connections and in this way merge to form a single unit: a multicellular organism. An early example of such an integration is the formation of plasma connection between normal cells and nitrogen-fixing cells in the blue-green algae. Later, clonally produced eukaryote cells became connected through plasma connections, the result being the multicellular eukaryote organisms. As was explained in SCIENCEBITES 1, 2 and 4 the criterion of plasma connections enables one to distinguish between a *pluricellular group*, which consists of cells that are attached without plasma connections and can have different genetic origins, and a *multicellular organism*, the cells of which are connected through plasma connections and are clonal (with rare exceptions, such as a chimaera, in which two zygotes have fused to produce one organism). Examples of pluricellular groups are the slug of a slime-mold, and a lichen. Examples of multicellular organisms are plants, fungi and animals.

Tick-marking Darwinian evolution. Using symbiogenetic mergers as a special kind of tick-mark, one can create a yard-stick for steps in Darwinian evolution. The idea is illustrated in the below pair of figures (Figure 7.2). The left figure shows Darwinian evolution without tick-marks. The color figure focuses on the tick-marks indicating the formation of a cell, the formation of a eukaryotic cell (with endosymbiont cells), and the formation of multicellularity (here only for eukaryotic cells). Using a figure like this as a guidance, one can explore questions about the kinds of the tick-marks, e.g.

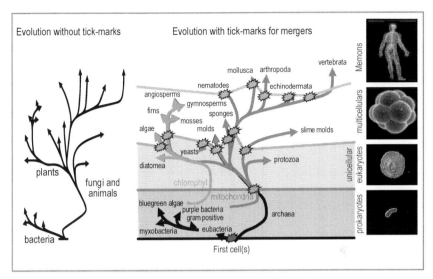

Figure 7.2. A comparison of the tree of life without (left) and with tick-marks for mergers (right).

are they of the 'right' kind, or of the 'same' kind, and are they located at the 'right' positions?

Natura non facit saltus. Does a merger, such as when several individually-living clonal cells form a multi-cellular organism, represent a gradual development or a leap? I think the answer depends on one's focus on either physical processes or on logical classes.

On the one hand, a focus on physical processes implies that any event depends on preparatory steps, on a range of attempts, on scaffolding conditions, etc. Such scaffolds prepare the way for the formation of a new merger. For example, in the mammalian embryo three generations of cell divisions generally precede the formation of plasma connections at the eight cell and later stages.

On the other hand, the French philosopher and naturalist Bonnet (in Arthur Lovejoy p. 275-276) already writes that a qualitative difference between two things necessary implies a discontinuity in a classification. With respect to levels of organization, this means that an organism is either unicellular or multicellular, but not both at the same time. The enclosure of one or more endosymbiont cells inside a host-cell, and the formation of plasma connections between cells, thus offer binary *logical* tick-marks for spitting up the continuous evolutionary series of *physical* forms. The tick-marks for logical classes are cell, eukaryote (with endosymbiont), multicellular organism, and eukaryote multicellular organism. And even though most biologists think of the brain as an organ of a multicellular organism, the operator theory adds as a next class all the organisms that have brains with dual closure. Mergers as discussed above thus represent states of matter that simultaneously represent a physical organization and a logical class.

Further reading

Darwin, C.R. (1859). On the origin of species by means of natural selection, or the preservation of favoured races in the struggle for life (1st Ed.). John Murray, London, UK.

Dennett, D. (1995). Darwin's dangerous idea: evolution and the meanings of life. Simon & Schuster, New York, NY, USA.

Jagers op Akkerhuis, G.A.J.M. (2016). Evolution and transitions in complexity. The science of hierarchical organization in nature. Springer, Cham, Switzerland.

Kozo-Polansky, B.M. (1924). Symbiogenesis: a new principle of evolution. Harvard University Press, Cambridge, MA, USA.

Lovejoy, A.O. (2009). The great chain of being. Transaction Publishers, Edison, NJ, USA.

Mereschkowski, K. (1905). Über Natur und Ursprung der Chromatophoren im Pflanzenreiche. Biologisches Centralblatt 25: 59-604.

8. Great chain of being

The great chain of being. In 1933 Arthur O. Lovejoy gives a series of 'William James lectures' at Harvard. The texts are published in his 1936 book titled: 'The great chain of being'. Lovejoy's book, with all the many interesting citations in it, forms the foundation for this SCIENCEBITE. The great chain offers a ranking of entities in the world, a 'ladder'. The idea of a ladder of nature (in Latin: *'scala naturae'*) dates back to Plato and Aristotle. In the middle ages (500-1500) Christians believed in a god who was the creator of all the entities of the great chain (Figure 8.1). The ranking of the chain was based on 'perfection'. God, angels and man resided at the top. Gottfried Wilhelm von Leibniz (1646-1716) suggested a non-religious version: *'Thus men are linked with the animals, these with the plants and these with the fossils, which in turn merge with those bodies which our senses and our imagination represent to us as absolutely inanimate'.*

Figure 8.1. The great chain of being (Didacus Valades, Rhetorica Christiana, 1579).

An old-fashioned idea? Today many scientists view the great chain of being as an old-fashioned idea. Their criticism is inspired e.g. by the following insights: (1) the classic chain is not based on mechanisms, but on some form of idealism; (2) gods and angels cannot be reconciled with a naturalistic approach; (3) the classification of the great chain is static. Stasis conflicts with current ideas about evolution.

Despite all these valid objections, it is difficult to deny that a multicellular plant is more complex than a bacterium, and a cell is more complex than a molecule. If a suitable mechanism were available, maybe it would be possible to identify and rank complexity in nature in a consistent way using levels after all. Have the scientists who condemn the great chain thrown the baby out with the bathwater? I will look for an answer to this question by analyzing historical revisions of the great chain, from a static chain, via a dynamic chain, to a dynamic chain with static tick-marks (see also SCIENCEBITE 7 about tick-marks).

Plenitude, continuity and gradation. Until the 18th century, the great chain is a static structure, based on entities that have fixed forms. It is associated with three philosophical concepts: plenitude, continuity and gradation. Plato is generally viewed as the inventor of the great chain. Plato's philosophy sketches a realm of fixed ideas. These ideas have a natural tendency to actualize themselves in the physical world. It is presumed that there is a 'plenitude' of ideas, causing a 'fullness' of actual things in the world, including the realization of all possible intermediate forms. In relation to this, Thomas Aquinas (1225-1274) writes: *'nature does not make [animal] kinds separate without making something intermediate between them; for nature does not pass from extreme to extreme* nisi per medium'. Later Leibniz (1646-1716) speaks about a 'law of continuity' through which *'all the orders of natural beings form but a single chain, in which the various classes, like so many rings, are so closely linked one to another that it is impossible for the senses or the imagination to determine precisely the point at which one ends and the next begins...'* Leibniz's message is: nature makes no leaps, while all physical forms are eternal, and linked through gradations of intermediates. In current science this 'law of continuity' reverberates in gradualist viewpoints.

Towards a dynamic great chain. During the 18th century, observations of change in nature make philosophers doubt the static great chain of being. Denis Diderot (1713-1748) writes that: '*At least it appears that Nature has never been, is not, and never will be stationary, or in a state of permanence; its form is necessarily transitory.*' People start asking whether species are created, or are produced by dynamic, causal processes. Early thoughts about a small number of ancestors being the source of all organisms of all species trace back to Pierre-Louis de Maupertuis (1745, 1751) and Diderot (1749, 1754). Roughly 100 years later Charles Darwin (1859) elaborates a theory for evolution based on '*descent with modification through variation and selection*'. In line with these developing insights, scientists no longer view the biological part of the great chain as an immutable structure. They adopt a dynamic view. Currently, hardly any biologist will seriously consider a static view suggesting the creation of species. Instead the focus has shifted towards a dynamic view in which all organisms are part of one large pedigree.

Continuous or discontinuous? In a text in which he defends the 'law of continuity', Robinet comments on the work of Bonnet, who in Robinet's view is too bold when he suggests to '*divide the different orders which constitute the scale of being into four general classes: (1) inorganic; (2) organic but inanimate (i.e. plants); (3) organic and animate, but without reason; and (4) organic, animate, and rational*'. In the eyes of Robinet, these four classes deny the law of continuity, because the qualitative differences that are used as criteria for the classes necessarily imply discontinuity. The classes suggest that a thing is either inorganic or organic, plant or animal. In contrast to Bonnet, Robinet is convinced that the great chain cannot be continuous and discontinuous at the same time. Can the opposing perspectives of Bonnet and Robinet be reconciled?

Physical continuity. In line with recent thinking, all the entities that constitute the great chain can be viewed as the products of a long series of processes in which past entities produced more recent entities. The production of a new physical or biotic entity always requires physical processes: existing physical entities combine to form a new entity, an existing entity splits, or an existing entity builds a new entity. Because of the necessity of such processes, nature cannot make leaps (at least not without scaffolding).

Logical discontinuity. In the great chain new physical entities either form from, or are constructed by pre-existing physical entities. The physics are continuous. Discontinuity, however, may enter the great chain as a logical illusion. This illusion starts with the naming of things. Names such as a 'molecule', or a 'cell', are a matter of human classification. The construction of a classification is a logical process. In logic, one needs criteria for deciding whether or not a thing fits into a specific class. In the ideal case, properties are available for assigning every entity to a specific class. To create a stringent connection between properties and classes, it is convenient if both the criteria for the classes, and the properties of the entities, are 'binary'. Binary means present or absent. Binary also means that when the criteria are met, the organization is of kind X, and when the criteria are not met, the organization is not of kind X. I have advocated above that in nature all physical processes are continuous. Can such continuity be reconciled with a binary classification?

Circularities. As an answer to the latter question, one can focus on 'circularity' or 'globularity' as a property of physical entities. In a logical sense, mathematical forms, such as a triangle, a square, a circle, or globe, can be used for classification purposes because they offer criteria for distinguishing one form from other forms. Interestingly, some organizational criteria exist that are to a large extent binary. This is typical for circular kinds of organization. For example, a piece of rope is either curved and open, or, when the ends are knotted together, circular and closed. This example shows that circular aspects of physical systems offer criteria for a binary classification. Circular properties of, and circular dynamics in physical systems thus offer a way to reconcile physical continuity with a binary classification.

Criteria for circularity. A strong set of criteria for a binary classification can be obtained by combining structural and functional circularity. For example, a cell has a membrane. This matches with structural circularity. And a cell has autocatalysis. This matches with functional circularity. In addition, and as a third criterion, one can demand interdependence; the membrane keeps the molecules of the autocatalytic set together, and the autocatalytic set produces not only its own molecules, but also those of the membrane. This combination of criteria is termed 'dual closure' in the operator theory.

Reconciliation. Starting with quarks, a sequence of entities has formed in nature possessing increasingly complicated kinds of dual closure: hadrons, atoms, molecules, cells, endosymbiont cells (the eukaryotes), multicellular organism, and multicellular organisms with brains. Each entity has its proper dual closure. Successive dual closures create a logical 'great chain' of entities with increasingly complicated organization. In my view, a 'ladder' based on dual closure offers a new way of reconciling the classical view of the great chain, with physical continuity of the world, and logical tick-marking of human classification.

Further reading

Darwin, C.R. (1859). On the origin of species by means of natural selection, or the preservation of favoured races in the struggle for life (1st Ed.). John Murray, London, UK.

Jagers op Akkerhuis, G.A.J.M. (2016). Evolution and transitions in complexity. The science of hierarchical organization in nature. Springer, Cham, Switzerland.

Lovejoy, A.O. (2009). The great chain of being. Transaction Publishers, Edison, NJ, USA.

Moreno, A., Mossio, M. (2015). Biological autonomy. A philosophical and theoretical enquiry. Springer, Dordrecht, the Netherlands.

9. A new view on evolutionary transitions

The logic of transitions. Changes in organization are sometimes named 'transitions'. There exist many kinds of transitions, e.g. from water to ice, from a living to a dead sheep, from a teenager to an adult, etc. Sometimes transitions can be ranked. For example one can transform metal into car parts, and transform car parts into a car. This SCIENCEBITE focuses on the consistency of such rankings. Assuming a constant logic for all steps in a ranking, the result can be viewed as being 'consistent'. I demonstrate that a focus on consistency adds new insights to the theory about 'major evolutionary transitions' in biology.

Logically consistent rankings. When is a ranking 'consistent'? An example if offered by the *natural* numbers: 1, 2, 3, 4, etc. Such numbers can be constructed if one starts with 1 and adds 1 every time. The resulting ranking has a consistent logic for two reasons: 1. The rule for creating a next number is always the same, 2. Every new entity produced by the rule is always of the kind 'natural number'.

A biological universe of discourse. A theoretical world with only natural numbers does not allow for all kinds of calculations. For example, if one subtracts 5 from 2, one obtains minus 3, which is a new *kind* of number. Minus 3 belongs to the so-called *integers*. In the world of the integers both addition and subtraction always produce another integer. But a fraction of two integers is not always an integer. For example, dividing 5 by two results in 2.5. This is not an integer. Only specific operations with integers result in new integers.

Because of such limitations, kinds of numbers can be viewed as specific worlds of things one can talk about, so-called 'universes of discourse'. By working in a specific universe of discourse, a mathematician knows exactly which activities make sense. In this SCIENCEBITE I explore the question of whether or not there exists a universe of discourse for biology? A biological universe of discourse could be constructed if a rule were available that, instead of producing numbers, produces and ranks higher complexity organisms. The quest for a biological ranking can profit from studying classical rankings of increasingly complex physical objects and/or organisms.

Classical rankings. Classical rankings that include physical and biological entities are e.g. the great chain of being (the *scala naturae*) of Aristotle, the approach of Pierre Teilhard de Chardin, and rankings by Arthur Young, by James Grier Miller, and by Allerd Stikker. In biology early rankings are suggested in the works of George Ledyard Stebbins, of John Tyler Bonner, and of Leo Buss. As a general illustration of how one can analyze the consistency of the logic of these and similar approaches I will focus on the recently published 'major evolutionary transitions'.

Major evolutionary transitions. In 1995 eight examples of so-called 'major evolutionary transitions' (MET) were published by John Maynard Smith and Eörs Szathmáry. Every major evolutionary transition can be viewed as a specific tick-mark on a yard-stick (as discussed in SCIENCEBITE 7). And one can study the consistency of the tick-marking process. I use the following two criteria for consistency: 1. Does each of the major evolutionary transitions (METs) follow the same rule, and 2. Does each major transition produce new entities of the same kind. If these criteria are not met, some MET's may – metaphorically speaking – indicate centimeters, while others indicate degrees Celsius, or kilograms, etc.

If the rules for producing entities and/or if the kinds of entities are different, it can be questioned from a logical perspective whether the transitions can be ranked along one and the same yardstick. Interestingly, it has always amazed me that the checking of kind-similarity has received little attention in biology. Things are very different in physics. Physicists consider it common practice to carry out a quality check of an equation by demanding that the equation has the same units on both sides.

Kind-similarity of the eight MET's. It is hard to say whether Szathmáry and Maynard Smith (1995) aimed at a stringent ranking of the major transitions they proposed. If such ranking is not intended, the order in the below summary of their MET's is to be understood as incidental. Potentially, however, the ranking may represent some order, and assuming this is true, I want to show that it is possible to gain new insights by analyzing the 'kind' of each MET. To find out about the kind, I discuss the eight MET's of Szathmáry and Maynard Smith individually:

1. Replicating molecules to populations of molecules in compartments: This can be viewed as a transition from chemical reactions to a cell.

2. Unlinked replicators to chromosomes: This is an event inside a cell that improves the cell's reproductive efficiency.

3. RNA as gene and enzyme to DNA and protein (genetic code): This represents a refinement of the cell's functioning.

4. Prokaryotes to eukaryotes: Here a bacterial cell turns into a more complex cell with endosymbionts and a nucleus.

5. Asexual clones to sexual populations: Sexual interactions offers a new mechanism allowing a grouping of organisms based on the *in-principle* possibility of gene transfer during mating.

6. Protists to animals, plants and fungi (cell differentiation): This is a transition from eukaryote unicellularity to eukaryote multicellularity.

7. Solitary individuals to colonies (non-reproductive castes): The formation of a colony from individuals is based on a behavioral grouping.

8. Primate societies to human societies (language): Language is used as a special behavioral grouping criterion.

Not one yard-stick, but three. The above discussion can be used as a basis for reorganizing the eight MET's into three distinct groups as follows:

(A) MET nr. 1, 4 and 6. In this group the transitions cause higher complexity organisms.

(B) MET nr. 2 and 3. In this group the transitions cause a more complex interior of an organism.

(C) MET nr. 5, 7 and 8. In this group the transitions are based on various kinds of interactions between organisms that remain separately countable individuals.

Each of these three kinds of transitions (A), (B) and (C) can be viewed as belonging to a different 'dimension' that has its proper rules for ranking.

In class (A) the cell, the endosymbiont cell, and the multicellular can be ranked in a stringent way as the products of successive dual closures (see e.g. SCIENCEBITE 1). The operator theory calls ranking along this dimension 'upward'. Along this dimension one can observe 'upward' transitions towards a higher number of preceding dual closures, and 'downward' transitions towards a lower number of preceding dual closures. For example if a plant produces single-celled reproductive cells, the level of dual closure changes 'downward' from multicellularity to that of a cell that hosts endosymbiotic cells in its interior.

In class (B) the transitions to chromosomes, and to DNA can be viewed as developments inside an organism. The operator theory refers to ranking along this dimension as being 'inward'.

In class (C) the transitions to populations, colonies, and society involve interactions between individual organisms. The operator theory refers to groupings along this dimension as 'outward'. In rare cases the interactions between organisms along the outward dimension become sufficiently 'close' to classify as a new kind of organism. The operator theory assumes that a new organism forms when, and only when, interactions scaffold the formation of dual closure.

Three dimensions for transitions. Above I discussed the integers, and the universe of discourse they form. The aim was to illustrate that a consistent ranking selectively includes entities that are identical in high-level kind, such as the integers. Like a ranking of integers, a ranking in biology must also be based on transitions that are identical in their high-level kind. To meet this criterion a ranking must be based on transitions of the same major kind, for example because every transition involves dual closure, and must include entities of the same major kind, for example because every transition produces a new, higher-level operator.

In relation to the work of Szathmáry and Maynard Smith, and similar approaches to transitions, it is relevant that it was demonstrated that the 8 major evolutionary transitions belong to three separate groups, each group containing entities of different kinds, while the ranking of entities in each group is linked to its proper, group-specific mechanisms.

Further reading

Bonner, J.T. (1974). On development: the biology of form. Harvard Univ. Press, Cambridge, MA, USA.

Buss, L.W. (1987). The evolution of individuality. Princeton University Press, Princeton, NJ, USA.

Jagers op Akkerhuis, G.A.J.M. (2016). Evolution and transitions in complexity. The science of hierarchical organization in nature. Springer, Cham, Switzerland.

Miller, J.G.M. (1978). Living Systems. Mc Graw-Hill, New York, NY, USA.

Stebbins, G. (1969). The basis of progressive evolution. University of North Carolina Press, Chapel Hill, NC, USA.

Stikker, A. (1992). The transformation factor. Towards and ecological consciousness. Element, Rockport, MA, USA.

Szathmáry, E., Maynard Smith, J. (1995). The major evolutionary transitions. Nature 374: 227-232.

Teilhard de Chardin, P. (1969). The future of man. Collins, London, UK.

Young, A.M. (1976). The reflexive universe. Delacorte Press, New York, NY, USA.

10. Phenotype and genotype of Big History

This SCIENCEBITE discusses the structuring of Big History. Big History is the name for a multidisciplinary approach that aims at a historical reconstruction, a timeline, of phenomena that since the Big Bang have occurred in the universe. Personally I encountered my first picture of the grand timeline of the universe in the book by Heinz Pagels (1985), where the long period of the history of the universe was divided into nine periods. I have since seen many similar timelines. Depending on the author, the historical timelines are ranked according to various principles. For example, Eric Chaisson uses epochs, while Fred Spier uses eras, or regimes. In this SCIENCEBITE I discuss different classifications. At the end I suggest an innovation to the classical analyses of Big History by distinguishing between a 'genotypical' perspective and a complementary 'phenotypical' perspective.

Epoch. Chaisson divides the history of the universe into eight epochs, termed: particulate, galactic, stellar, planetary, chemical, biological, cultural, and future. Every epoch is characterized by a dominant phenomenon. The particulate epoch is associated with 'fundamental particles'. The stellar epoch is associated with stars. The chemical epoch is associated with molecules, and the biological epoch with organisms. Next, a switch is made to culture, which refers to learned behavior. Clearly, culture is not a material 'thing' but an indication of special behavioral patterns. The last epoch is 'the future', which is an imagined period. These examples demonstrate that the entities used for defining the different epochs belong to different logical kinds.

Era. Spier speaks about eras when referring to the Radiation Era, or the Matter Era. In science, eras are generally used for long periods during which specific phenomena occur. Eras are subdivisions of still longer periods, called eons. In the last 543 million years geologists distinguish three eras that are associated with the existence of organisms: the paleozoicum (oldest organisms), the mesozoicum, and the kenozoicum.

Discussing epochs and eras. Both the concepts of epoch and era indicate a period during which a specific phenomenon can be observed. Such phenomena may differ in kind. Some epochs involve *objects*, such as fundamental particles, stars, or organisms. Other epochs involve *abstractions*, such as radiation, matter, culture or 'the future'. While everyone is free to define an individual epoch or era to his/her liking, and in relation to any kinds of entities, I advocate that for a ranking to be logically consistent, it must include entities of the same major kind across its entire range. For example, one can focus on celestial entities, such as matter clouds, black holes, galaxies, stars, planets, moons and comets. Or one can focus on material 'particles', such as quarks, hadrons, atoms, molecules, cells, etc. In Big History, the ranking of epochs or era's poses a challenge, because the development of the universe can be looked at from different perspectives, each resulting in a focus on different entities, and in different rankings.

Regime. Spier uses the term 'regime' as the cornerstone of his analysis of Big History. The term regime is chosen because of its malleability, allowing a broad use. The term refers to a *'more or less regular, but ultimately instable process-structure'*. Basically, Spier distinguishes two kinds of organizational regimes: complex adaptive systems, and complex non-adaptive systems. Examples offered of non-adaptive regimes are: stars, galaxies, black holes. Regimes can vary in size from celestial regimes to the regimes of the tiniest particles. Spier also defines life as a regime that uses a hereditary program for directing mechanisms that extract energy from the environment and allow maintenance and – if possible – reproduction.

Discussing regimes. The term regime is commonly used in the context of hydrology, where the 'water regime' of a river refers to the range in the river's yearly discharges. And climate can have a 'regime shift' when the fluctuations of yearly temperatures change. A society may shift from a democratic regime to a dictatorial regime. In all such examples the term regime refers to boundary values of dynamics and/or organizational properties of a system.

This also shows that the term *regime* and *system* define different things. To prevent ambiguity on this point, I suggest using the following definition of the term *system*: *a part of the universe that is analyzed in a systemic way*.

A systemic analysis implies that one focuses on parts of the system, and their relationships. Relationships may be static (e.g. distance, difference) or dynamic (e.g. motion, interaction). The system can be of any kind, e.g. an atom, a bacterium, a galaxy, a car, a society, or a government, and can be static, e.g. when frozen, or dynamic, e.g. the development of a city. The system definition offered here applies to any part of the universe and can help resolve ambiguity about the system concept. With the help of this definition, one can analyze any and all properties of a system.

Meanwhile, the concept of a regime suggests a special focus. For example, a river, as a system, can have a water regime. And an iron atom, as a system, has a more or less fixed organization (a regime of organization, or *kind* of organization) consisting of a nucleus which is surrounded most of the time by 26 electrons.

Operators and interaction systems. Rankings based on epochs, eras, or regimes don't automatically discriminate between particles, celestial bodies, behaviors, etc. The inability to discriminate kinds of entities implies that one runs the risk of creating a mixed ranking, because one cannot base the entire ranking on one specific kind of transitions between classes. As a solution, I suggest thinking along two lanes by differentiating between the rankings of two major kinds of entities. The first kind are the operators. The second kind are systems that consist of interacting operators without being an operator. The operator theory refers to the latter systems as 'interaction systems'. Until now, and starting with quarks, operators of the following kinds have formed: hadron, atom, molecule, cell, endosymbiont cell (the eukaryote cell), multicellular, endosymbiont multicellular, and neural network organism (see also SCIENCEBITES 4, 8, 9). Interaction systems in the universe are many, including stars, planets, black holes, galaxies, etc.

Phenotype and genotype of Big History. Both the operators and the interaction systems have played their roles during the formation of the universe. It is a challenge for Big Historians to address these intertwined roles (Figure 10.1), while still guaranteeing the logical consistency of a ranking. Classically, Big History focuses on the aggregation of clouds of particulate matter into galaxies, the collapse of matter to black holes, the formation of

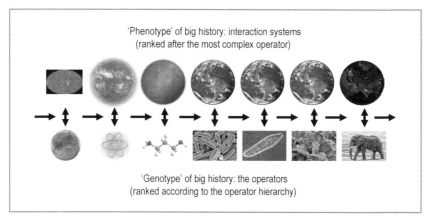

'Phenotype' of big history: interaction systems
(ranked after the most complex operator)

'Genotype' of big history: the operators
(ranked according to the operator hierarchy)

Figure 10.1. The link between the 'phenotypic' ranking of Big History based on celestial bodies and the 'genotypic' ranking of Big History based on the operator hierarchy.

stars, the explosion of super nova, the formation of planets, and moons, and the formation of ecosystems and culture.

I suggest calling this focus on observable, larger things, the 'phenotypic' approach of Big History. Using the analogy with the phenotype and genotype of an organism, the phenotype of Big History has its complement in a 'genotype' that is based on the stringent ranking of the operators, from quarks to neural network organisms. The ranking of the operators can be viewed as the 'informational' side of Big History.

During the development of the universe, the genotypic and phenotypic aspects remain linked in all aspects. For example, stars, as phenotypes, are the interaction systems in which new operators, such as atoms and molecules form. In turn, the formation of new operators will affect any further development of the interaction system in which they formed.

Wrapping up. Using the hierarchy of the operators one can create a stringent ranking of epochs and eras. For example, as soon as the first cells form, the universe changes from the molecule era to the cell era. As long as at least one cell exists somewhere in the universe, the cell era continues. And a planet can be classified after the highest-level operator, for example as a planet that is inhabited by unicellulars, or one inhabited by multicellulars. These examples illustrate how the operator theory suggests innovative 'genotypic' and 'phenotypic' viewpoints for rankings in Big History.

Further reading

Anonymous (2016). Big history. Dorling Kindersley Limited, London, UK.

Bryson, B. (2003). A short history of nearly everything. Transworld Publishers. London, UK.

Chaisson, E.J. (2001). Cosmic evolution: the rise of complexity in nature. Harvard University Press, Cambridge-London, UK.

Jagers op Akkerhuis, G.A.J.M. (2016). Evolution and transitions in complexity. The science of hierarchical organization in nature. Springer, Cham, Switzerland.

Pagels, H.R. (1985). Perfect symmetry. Simon and Schuster, New York, NY, USA.

Spier, F. (2010). Big history and the future of humanity. Wiley-Blackwell, Chichester, UK.

11. The species delusion

Do biologists need species? Historically, the species concept has seen marked changes in popularity. Plato's eternal ideas, Aristoteles' great chain of being, and God's creation offered inspiration for fixed species. A more recent classical view, the 'principle of continuity' favors a connected gradient of individual forms. Charles Darwin's evolution theory, and the modern synthesis, basically use individuals to dynamically explain speciation. Are species fixed, or do they evolve? Or is the species concept a persistent delusion?

Fixed forms. The idea of fixed species is old. Plato envisioned an abstract realm with ideas that materialize in all the different objects in the world. Plato's ideas are fixed, eternal, and so are the objects in which they materialize. Aristotle, while rejecting fixed ideas, introduced another fixed system, the *scala naturae* (see SCIENCEBITE 8). The *scala* can be understood as a ranking that from top to bottom includes holy and supranatural things, humans, animals, plants, and lifeless matter (e.g. clay, stone). Fixed species have been criticized by philosophers, for various reasons, notably because species are man-made, and because the so-called 'principle of continuity' casts doubt on classifications that make use of stringent limits to any class, such as a species.

Continuity calls for individuals. The principle of continuity asserts that if God created an orderly arranged world with fixed forms, all the different kinds of things must be part of a continuous series. In a continuous series, all forms have intermediate forms. A continuous series makes it hard to delimit a species. As can be found in Arthur O. Lovejoy (2009), this is the reason that Georges-Louis Buffon (1749) concludes that the notion of a species must be artificial: '*In general, the more one increases the number of one's divisions, in the case of the products of nature, the nearer one comes to the truth; since in reality individuals alone exist in nature.*'

Charles Bonnet (1769) holds a similar view (in Lovejoy, 2009): '*If there are no cleavages in nature, it is evident that our classifications are not hers.*

Those which we form are purely nominal, and we should regard them as means relative to our needs and to the limitations of our knowledge. Intelligences higher than ours perhaps recognize between two individuals which we place in the same species more varieties than we discover between two individuals of widely separate genera. Thus these intelligences see in the scale of our world as many steps as there are individuals.' Interestingly, the principle of continuity leads to a focus on individuals, and the rejection of species. This leads to the question whether or not species are real physical entities, or just man-made categories?

Species as man-made categories. Species limits are contested. Continuity makes species limits melt. And species limits are distrusted because they are man-made. Oliver Goldsmith (1763, in Lovejoy, 2009) states that *'All divisions among the objects of nature are perfectly arbitrary',* and continues saying that *'...all such divisions as are made among the inhabitants of this globe,..., are the work, not of nature, but of ourselves.'*

About one hundred years later, Darwin (1859) discusses the evolutionary divergence of species while emphasizing that limits to species are imaginary:

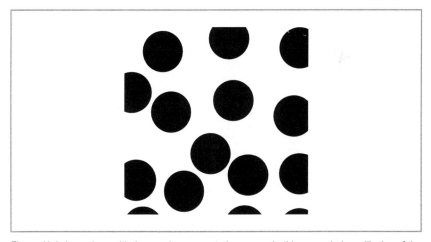

Figure 11.1. In analogy with the species concept, the square in this example is an illusion of the mind. The square is conceptual. In reality there are only dots with different sizes and shapes on a piece of paper. Similarly, instead of a population, there exist only organisms of different size and life-stage.

'In our diagram the line of succession is broken at regular intervals by small numbered letters marking the successive forms which have become sufficiently distinct to be recorded as varieties. But these breaks are imaginary, and might have been inserted anywhere, after intervals long enough to allow the accumulation of a considerable amount of divergent variation.' Such citations suggest that both Goldsmith and Darwin are aware that species are man-made groupings which in their views are arbitrary and imaginary (Figure 11.1).

Species and evolution. In a dynamic worldview, the branching of species is a conceptual tool for creating a pattern of evolution at a high level of abstraction. As part of this branching perspective, a (sexual) species is a 'group of individuals' that after many generations has 'branched' from an original group. A practical take on the species concept is that a species is a grouping of 'similar looking' individuals. But what is 'similar looking'? Must males and females look the same? Must different stages of the life history look the same? And what if individuals of different species look the same?

Because of the latter difficulties, biologists generally use more fundamental criteria for distinguishing a new species: (1) the individuals of the new species share a common descent with those of an old species; and (2) mating – when it occurs – causes fertile offspring only with individuals that belong to the species. These improved criteria are again not entirely satisfying. One reason is that in practice one rarely has certainty about descent (unless one carries out a detailed DNA study). And, even though it is normally quite easy to prove that an organism mates successfully with one or more neighbors, it is only rarely possible in a practical situation to test the mating success of all potential partners in the population. An even more difficult task is to carry out mating tests with the aim of creating conclusive evidence for the impossibility of species crosses, simply because this involves an even broader group of organisms. Meanwhile, a single species cross, however rare, may debunk the species limit. It is also relevant that if a definition makes use of mating, it will run into trouble when individuals are born infertile, and cannot mate 'within' the species. In such cases, descent is the sole criterion that can be used. All such practical aspects make it hard to empirically prove the two above species criteria for all the individuals that may belong, or may not belong to a species.

Induction. As a solution to the above empirical difficulties, one may accept that the species concept is only valid by induction. The assumption is that the production of fertile offspring by some parents is potentially valid for all other individuals with similar appearance when in the adult stage. Inductions are prone to Carl Popper's example of a 'black swan'; there can always be a male/female that does not want to mate, does not produce fertile offspring, or does produce fertile offspring with a partner of another species (e.g. the coywolf), etc. It is also problematic in practice that if one demands that offspring must be fertile, and one would like to check this criterion, one can only determine the membership of a species experimentally if one has access to several generations.

An 'in principle' definition. Maybe it is better to view the species concept as an 'in principle' definition. It is hard to empirically test descent, breeding and fertility of the offspring for every member of the species. And it is simply unthinkable to check the possibilities for cross-breeding with the organisms of any and all genetically 'closely' related species. The best approximation is to take samples and test for rates of successful mating. These considerations imply that an exact, and empirically testable species definition may well be a delusion. And it may not be the only delusion about species. Other delusions are uniformity of a species, and the species as a physical thing.

The uniformity delusion. Imagine a male animal that searches for a mating partner. One day, he discovers another organism in a nearby garden, a female, that he finds attractive. They mate. Some weeks later three offspring are born. One offspring catches a fever and dies. Later that year, the two remaining offspring migrate to neighboring gardens. This story can be told without a single reference to the species concept. Yet it captures all the ingredients for evolution: mating choice, offspring production, differential survival of the offspring in relation to their environment, migration.

Everything that happens to the above organisms occurs locally. Environmental opportunities and pressures are local. Genetic variation is local. Potentially, many other similar organisms exist in other countries, where they live different lives, experience different selection pressures, find local mates, and share local genes. The lives of the above organisms are connected to a few neighbors, while the connection with individuals in the population at large, the 'species', is indirect, through the spatial spreading of mating interactions and migration, over many generations and in a very large time frame. Because of genetic reshuffling during sexual reproduction, and the dominance of local interactions, the individuals of a species necessarily vary.

The thing delusion. As indicated above, the male and female animal discussed above know little about peers in other gardens, and are unaware of peers in other cities, or countries. They have no idea about something like a species range, a gene pool, or distortions of a Hardy Weinberg distribution. The latter are all conceptual terms used by humans to facilitate scientific communication, and for carrying out calculations. As I advocated in SCIENCEBITE 5, a species cannot drink from a pool. The physical entities drinking from a pool are the organisms. Likewise species cannot compete, or die. The moment the last individual of a species dies we say that the 'species' has gone extinct, but this is metaphorical language. A species is not a physical thing. It is a concept that is used to group 'similar' organisms that are potentially capable of mating and producing fertile, and sometimes infertile, offspring. Mental groupings, such as a population, a species, and a gene pool, have no role to play in the physical world. They are approximative tools in our models.

The species delusion. If the species is in many ways a delusion, it may be an interesting experiment to eliminate the term from biology. I consider this experiment especially relevant in the following situations: (1) when a species is used as an entity that can act in the field; and (2) when a species is used as a level in a constructive hierarchy in which organisms in interaction form species, while interacting species form communities. Such hierarchies exist only in our mind, not in the field. In nature, there is only one 'flat' world, in which individual biotic and abiotic objects interact with other objects.

In conclusion. Wrapping up, I argue that one must not think of a species as a physical thing (e.g. a species drinking from a pool). And if one uses the 'species' concept as an excuse for treating all organisms of a species as equal, and as spatially well-mixed, this looks more like chemistry than modern biology. In conclusion, from a physical point of view, the species is a delusion.

Further reading

Buffon, G.-L. (1749). L'histoire naturelle, générale et particulière, avec la description du Cabinet du Roi. Imprimerie royale, Paris, France.

Bonnet, C. (1769). Contemplation de la Nature (2nd Ed.). Marc-Michel Rey, Amsterdam, the Netherlands.

Darwin, C.R. (1859). On the origin of species by means of natural selection, or the preservation of favoured races in the struggle for life (1st Ed.). John Murray, London, UK.

Goldsmith (1763). A new and accurate system of natural history. The Monthly Review XXIX: 283-284.

Jagers op Akkerhuis, G.A.J.M. (2016). Evolution and transitions in complexity. The science of hierarchical organization in nature. Springer, Cham, Switzerland.

Lovejoy, A.O. (2009). The great chain of being. Transaction Publishers, Edison, NJ, USA.

12. The construction of time

This SCIENCEBITE is inspired by Carlo Rovelli's book 'L'ordine del tempo'. A marvelous book. Rovelli looks at time as a physicist. In this SCIENCEBITE I explore time as a physics-oriented biologist and philosopher.

Questions about time. Our everyday lives are deeply entrenched in time. We use time to weave the fabric of our planning. We make appointments using calendars, watches and mobile phones. We see indications for an arrow of time in how the world changes, and in aging. In science fiction novels the characters travel through time. But what does that mean? And why do people primarily associate time travel with exotic physics, instead of with biology? What is that thing called time? In this SCIENCEBITE I discuss such questions. As a foundation for discussing time, I start with three thought experiments.

Changeless existence. Imagine a world where nothing changes. Imagine every molecule, every atom, even every quark and photon in this world to be fixed at a precise location. Everything lingers in an absolute standstill. Such a world would look like a 3D picture. Even after millions of years things in this world would still be the same, in the exact sense of the word. Without change, a world like this exists in an eternal and absolute now. Since there is only now, there is no time. This world is timeless. Forever.

Timeless change. Imagine a different world, where everything is dynamic. Sadly, late-stage dementia has erased your memory. As a consequence you are aware of what happens, almost exactly when you observe it. You live in a here-and-now that extends indefinitely. You no longer understand change, causality, and what it means to speak about past or future moments. You experience an eternal 'now'.

Old jeans. Imagine your old jeans lying on the floor. What does 'old' mean? Your jeans look bleached, dirty. The cloth is thin. This is the 'state' of your jeans. You can describe the state of your jeans without knowing its past. The state of your jeans is a momentary observation. Next, you can link the state of your jeans to the six years of tear and wear since you bought them. Now you use memories to make a causal deduction of how your jeans changed over the past six years. Your jeans' age is linked to memories. When we speak about time, we have thoughts about experienced and imagined situations. In this web of thoughts, the present is just a thin veil in between past states that no longer exist, and future states that we imagine and that will possibly exist.

Constructing time. Above I introduce three notions relevant for discussing time: change, state, and memory. In the following lines I argue that none of these three topics by itself represents time. First, a thing can change. Change refers to a comparison of two states that occurred in a dynamic system, and the observation that the two states are different. The observation of such a difference is a difference, but difference is not time. Second, the state of a thing refers to how it 'is'. Not to time. Third, you can think about past or future experiences, and rank them. But such ranking represents order, not time, at least not in a quantitative sense. So what is time for us humans?

The arrow of time. Some people relate time to motion and the new system states resulting from it. For example, gas molecules in a container will diffuse from an unequal to an equal distribution. This is because the molecules move. Moving molecules experience less collisions in the direction of open space, which is why on average they move further in that direction. With time passing by this will cause the gas to spread out equally in the container. Motion and interactions are the causes of directional change. In the context of thermodynamics, people associate directional change with the 'arrow of time'. The arrow flies in one direction because autonomous dynamics are always linked to the dispersion of energy through the system, which implies a decrease in free energy.

The nature of time. I deliberately separate the discussion about the *arrow* of time and its direction, from the discussion of what time 'is', thus the discussion about the *nature* of time. The reason? Thermodynamics is about changes in the organization of the system. And autonomous changes are associated with an increase in the number of microstates of the system at large, which implies the dispersal of matter-energy. One can analyze a change in microstates by comparing a next state with a preceding state. This comparison can be done without a notion of time. The current discussion, however, is about how one could relate a succession of microstates to time. This means that the focus is on the nature of time.

A yardstick. I already suggested that our perception of 'change' depends on the ranking of memories. Our brain allows thoughts about past, current, and future states. In this context, I view 'time' as a measure that connects thoughts along a *theoretical, mental* yardstick. In order to quantitatively organize this mental yard-stick, people need tick-marks.

Tick-marking time. As a basis for quantitative time measurement, people need a yard-stick with tick-marks at equi-temporal distances. Such distances can be derived from a system that naturally performs a regular oscillation, which allows people to time-stamp an event, such that the moment when it happened can be checked later. On our yardstick for time, 24 hourly tick-marks can be inscribed by making use of the large oscillator we all live on, the earth, which rotates once every day. More rapid tick-marks can be derived from small oscillators, such as the pendulum of a mechanical clock. And at the very low end of the scale we find tick-marks as small as the oscillations of atoms in an atom clock. Currently, Ytterbium clocks are the most precise. By averaging over the oscillations of several clocks, scientists can produce a time measurement that is so accurate that the result would vary by less than one second over the entire age of the universe.

The use of standardized clocks and watches, which need not be as accurate as Ytterbium clocks, allows two people to make an appointment for a meeting at a bar at a moment in a not yet existing future, e.g. tomorrow, at 12.36. The story of the yard-stick and of timestamping holds for the world of large things that we all live in. I learned from Rovelli that on the smallest scale time exists as quanta. When time exists as quanta, it is no longer possible to create a yardstick for time.

Relativity of time. In addition to the diurnal cycle of dark and light, organisms have evolved their own physiological clocks. An organism becomes hungry or sleepy at more or less regular intervals, the intervals depending on internal physiological processes. In addition, intelligent animals can experience mental sensations of time. Both physiological clocks and mental time can easily be distorted (Figure 12.1). Desiccation and freezing arrest physiological time. Sleep, the use of drugs, drinking alcohol, and being absorbed in an activity distort mental time. And Adrian Bejan indicates that the older we get, the less time there seems to be in between years, because the number of impressions we process reduces slowly with age.

Our mental time is so variable, that we gladly rely on physical clocks as a reference. Physical clocks, however, can be distorted too, for example near the speed of light. The reason for this is that the speed of light is the maximum speed in the universe. If a thing traveled at the speed of light, its parts wouldn't move anymore, because any additional movement would imply that their speed would become faster than the speed of light. As a consequence, physical time comes to a standstill at the speed of light. In an analogous way strong gravity slows down time. The consequence is that in the universe every object has its proper pace, depending on local speed and gravity. This is the reason why physical time is said to be relative.

Figure 12.1. Both physiological clocks and mental time can be distorted.

Time-loops. Scientists analyze time as something that proceeds first A, then B, then C, etc. But what happens when nature is trapped in a cyclical interaction? For example, A, then B, then C, then A. Let's give this cyclical system from A to A the name 'X'. Interactions A, B, and C occur inside X while the dynamic cycle persists 'through' time. The persistence of X introduces a physical entity with a new time dimension: the existence of X has become invariant to lower level change. X has its proper lifetime. Dynamic cyclicity is a characteristic of a special kind of persisting objects that are called loops. And of all possible things with loops in the physical world, the 'operators' form a special subset, based on dual closure. Using quarks as the basis, the following operators have formed in our universe: the hadron, the atom, the molecule, the cell, the endosymbiont cell, the multicellular and the neural network organism. Every operator, of every kind, has its proper life-time.

Time travel. You can 'time-travel' when the parts that you consist of, e.g. atoms and molecules, move at a different rate than those of the world around you. For example, if you travel near the speed of light, the movement of your internal parts will slow down, as will the rate of your physiology, and your physiological time. Imagine that after many years of traveling at very high speed, you return to your original location. From your perspective of relatively slow time, it seems as if you have arrived in the future.

Compared to time travel in physics, which is a rather exotic phenomenon because of the extreme conditions that are required, time travel in biology is everyday practice. Millions and millions of seeds time-travel every year in the soil from autumn to the next spring. Likewise, squirrels, hedgehogs, and bears sleep their way through the winter. Dry eggs of tiny *Gammarus* crustaceans arrest their metabolism, and time, through desiccation. They lay for many years in the hot sand of the desert, waiting for heavy rainfall to fill the puddles and ponds where they can hatch, and develop to larvae that feed on algae, grow and reproduce.

A subtle way of time travel is a low calorie diet. Such a diet slows down your physiology, which lets you grow older. In the future, technical organisms (intelligent robots) will use time travel as a means to explore deep space. Robots can sleep for thousands of years, re-activating their technical brains upon arrival at their destination.

Time: a virtual measure. In the above I discuss aspects of time that offer support for the viewpoint that time is a virtual measure people use for organizing their thoughts about the past, the present and the future. One reason for using time is that it helps us to predict temporal patterns. Predicting such patterns enables survival in a complex world. Viewing time as something physical, e.g. something that a person can travel through, or something that can pass, is a philosophical category mistake. States and motion are physical and exist only in the here and now. Change and time are logical.

Further reading

Bejan, A. (in press). Physics explains why time passes faster as you age. European Review.

Biba, E. (2010). What is time? One physicist hunts for the ultimate theory. Available at: https://www.wired.com/2010/02/what-is-time/.

Jagers op Akkerhuis, G.A.J.M. (2016). Evolution and transitions in complexity. The science of hierarchical organization in nature. Springer, Cham, Switzerland.

Rovelli, C. (2017). L'ordine del tempo [in Dutch]. Prometheus, Amsterdam, the Netherlands.

Witrow, G.J. (1960). The natural philosophy of time. Oxford University Press, Oxford, UK.

13. One ring to rule them all

Lord of the rings. In his epic novel, John Ronald Reuel Tolkien writes: '*One Ring to rule them all. One Ring to find them. One Ring to bring them all and in the darkness bind them.*' This powerful poem encapsulates the central theme of the book. Tolkien's vision of a ring reverberates in the circular process of closure that, according to the operator theory, is fundamental to complexity in nature. This SCIENCEBITE tells the story of the discovery and impact of closure and its relevancy for fundamental concepts such as 'self', 'information', and 'unity'.

Hierarchy of levels. In 1992 I was offered the opportunity to write an integration study about the results of 22 PhD theses in terrestrial ecotoxicology. A major goal was to connect toxicant effects 'across levels of organization in the ecosystem'. Connecting effects across levels was a challenge because scientists at that time appeared to think and write differently about what is a 'level'. The common approach was that, in one way or another, entity A at level X is a part of the organization of entity B at level X+1. The world was compared with a nested set of Russian dolls.

In line with this, the famous biologist Ernst Mayr wrote: '*The complexity of living systems exists at every hierarchical level, from the nucleus, to the cell, to any organ system (kidney, liver, brain), to the individual, to the species, to the ecosystem, the society.*' A ranking like this involves different kinds of entities (see also SCIENCEBITE 1). A cell can be an independently dwelling entity or a part of an organism. An organ is a part of a multicellular organism. The term individual can refer to a single countable entity of whatever kind. A species is a grouping of organisms. An ecosystem is a grouping of organisms and abiotic objects. Moreover, the entities at the successive levels of Mayr are connected through different kinds of relationships, e.g. physical interaction, conceptual grouping, etc.

Theoretically it would be an advantage if entities at different levels would all be of a similar major kind, e.g. 'physical objects', and that clear rules would describe the nature of the entities. For example from molecule to cell, one can propose that 'lower level entities interact in a circular way to construct the higher-level entity'. And from organism to population, one can propose that 'some kind of similarity is used for the conceptual grouping of selected organisms into a set called the population'. If throughout a ranking entities and rules are of the same kind, this makes the ranking logically consistent in this respect. When developing theory, I consider logical consistency an important criterion for scientists to strive for.

Layers of complexity. A logically stringent ranking demands that one has access to criteria that allow the identification of a lower and upper boundary to every 'level' or 'layer'. To demonstrate how one can work towards such boundaries I use the atom as an example. Every atom has a nucleus surrounded by electrons. One can say that the combination of a nucleus and electron shell define the atom. One can proceed from this basis, by linking atoms together through covalent bonding. Covalent bonds result in units that consist of two or more atoms, called molecules. The atom and the molecule define the lower and upper limit of a 'layer' that harbors things that are atoms, or are based on atoms. By analogy, a bacterium, a blue-green alga, a protozoa, and a plant are all based on cells. The cell and the multicellular organism provide a lower and upper boundary of the layer of things that are cells, or are based on cells. This way of analysis offers a first approximation of lower and upper limits as a basis for a stringent layering. Next, precise criteria are needed for every kind of object involved, e.g. atom, molecule, cell, eukaryotic cell, etc. On this point, Stuart Kauffman's book 'The origins of order' inspired me.

Functional closure. While reading Kauffman I was thrilled by the Boolean networks, and how such networks can produce a closed chain of interactions: a 'loop', or in my own words, a 'closure'. In a catalytic Boolean network, closure occurs when reactant A catalyzes the production of reactant B, which catalyzes the production of C, etc., until an Nth reactant produces A, and the loop closes.

Closure allows for a binary distinction between the state of the system before closure, and the state of the system after closure. The magic that happens is that closure *both functionally and analytically* unites a set of catalytic reactions into *set-wise* autocatalysis. The closing of a loop of catalytic transformations thus offers a stringent criterion for a stepwise, and causal, analysis of organization.

However, there are two additional subjects that also need to be dealt with. The first is that without a containing structure, the catalytic molecules can diffuse in all directions, and their closure will be lost. The second is that when the molecules of two autocatalytic sets float in the same environment, they can mix freely. Accordingly one cannot identify one from the other. Because of these two phenomena one cannot say that an autocatalytic network as such represents a single countable *physical* unity. Additional properties are needed.

Structural closure. To physically unite all the molecules of an autocatalytic set that is dissolved in water, the molecules must be kept together by a material structure. Such a structure creates 'physical closure'. As an example, one can think of the bacterial membrane confining the molecules of the autocatalytic set to a limited space. Of course, not all molecules in the cell need to be confined, because to maintain the autocatalytic processes, 'food' molecules must enter, and 'waste' molecules must leave the 'closed' environment. The membrane only has to 'mediate' the molecules of the autocatalytic set.

Dual closure. In more general terms, the autocatalytic set and the membrane allow for a 'functional' closure and a 'structural' closure, respectively. Moreover, the functional and structural closure are mutually dependent. For example, in the cell, the autocatalytic set is contained by the membrane, and the material of the membrane is produced by the autocatalytic set. I named such mutual dependency 'dual closure'. For me, dual closure represents the organizational equivalent of the ring of power of Tolkien, because, like with so-called 'Borromean rings', the three rings exist separately, but are necessarily attached (Figure 13.1). It must be stressed that the name dual closure offers only a minimal conceptualization, as it focuses on the two most relevant rings represented by structural and functional closure. In a broader context any dual closure can be viewed as to be direct successor of the dual closure defining the level below. Looking at the bacterial cell as an example, both the autocatalytic set and the membrane are based on molecules.

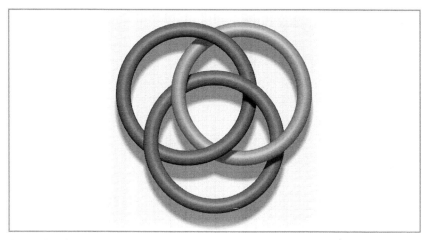

Figure 13.1. Borromean rings. Typical about these rings is that the linkage of every ring depends entirely on the presence of the other two rings. If only a single ring is taken out, the coupling of the rings is lost. This idea can serve as an analogy for dual closure.

At this point, it is relevant to explain that the occurrence of a next dual closure is more important than its construction from the operators at the preceding level. Whilst from the quarks up to multicellular organisms every closure step depends for its construction on interactions between preceding level operators, this logic is not leading for the identification of dual closure. To explain this, the brain forms an illustrative exception, because animal's brains are not constructed from multicellular organisms. What is important for the logic of dual closure when it is applied to the brain is that dual closure relates to the preceding highest level of organization, the multicellular environment of a multicellular organism, and that groups of cells in this environment interact in a way that already allows for dual closure, such that there is no necessity of dual closure having to be based on interacting organisms. By analogy, a future intelligent robot organism need not be constructed from animals. Instead intelligent beings can construct a robot having dual closure.

Now that some major aspects of dual closure have been introduced, I will use the concept as a context for discussing some fundamental scientific concepts such as 'self', 'information' and 'unity'.

Dual closure causes a self. The term 'self' is an enigmatic concept. It pops up in the word auto-catalysis (the Greek word 'autos' means self), and in terms such as self-organization, and self-awareness. In principle the term 'self' refers to an entity '*itself*, which refers in a holistic way to the entity. Not every use of 'self' is the same. If a molecule A transforms another molecule X into a second molecule A, the first A has not literally recreated itself, but has created a new molecule A that is structurally identical. Viewed this way, a molecule is not capable of self-reproduction. Things are different for the sets of entities that maintain dual closure. For example, self-maintenance of a cell implies the continuation of the dual closure of that cell as the fundamental aspect of its self. From the viewpoint of dual closure it becomes irrelevant whether it is molecule A or a copy of molecule A that is involved, as long as all the individual catalytic reactions assure that the overall dual closure remains intact.

Dual closure as a context for information. Information has many meanings. The Roman term 'informare' means 'to put into shape'. In the Roman sense, any structure is a thing that has a specific shape, and that represents information for this reason. Such a perspective was used by Claude Elwood Shannon and Warren Weaver when they referred to a structure with substructures that each minimally code for yes or no.

Related to this approach is a focus on semantic information, as has been advocated e.g. by Peter Checkland and Jim Scholes. Now information applies to data that can be endowed a 'meaning' in a specific context (e.g.). The term '*meaning*' generally refers to a change in uncertainty in a decision process. For example, two lovers can agree that one will send the other an empty envelope to confirm their marriage. An empty envelope contains neither a letter, nor written information. Yet, the envelope offers contextual information, the context being the agreement between the two lovers.

In line with the semantic interpretation, the cyclical relationships involved in dual closure create a context for the informational value of the structures creating and supporting these cycles. The context is the continuation of the dual closure. For example, in the cell, every catalyst has its context in the overall maintenance of the cell. While catalytic molecules represent 'data', it is their role in the cell that turns their structure into information.

Dual closure defines unity. Many things can be viewed as units. Of all possible things, the objects with dual closure, the operators, form a very special subset. The operators are special because the unity of an operator is defined in a triple way. 1. Unity because of functional closure. 2. Unity because of structural closure (which makes the operator a single physical object). 3. Unity because the structural and the functional closure are mutually dependent in an obligatory way.

Three rings to rule them all. When Tolkien wrote: *'One Ring to bring them all and in the darkness bind them'* he did not think of closure, nor of the operator theory. Yet, one can rephrase Tolkien's sentence to 'Dual closure to rule them all and in cooperation bind them'. Now Tolkien's epic poem becomes a reference to the relevancy of dual closure in system science, biology and philosophy. Dual closure binds entities into units, and in this way gives meaning to terms such as self, information and unity.

Further reading

Checkland, P., Scholes, J. (1990). Soft systems methodology in action. John Wiley & Sons, Chichester, UK.

Hofstadter, D. (2007). I am a strange loop. Basic books, New York, NY, USA.

Jagers op Akkerhuis, G.A.J.M. (2010). The operator hierarchy. A chain of closures linking matter, life and artificial intelligence. Alterra scientific contributions 34. Wageningen Environmental Research, Wageningen, the Netherlands.

Kauffman, S. (1993). The origins of order: self organization and selection in evolution. Oxford University Press, Oxford, UK.

Mayr, E. (1988). Essay one: is biology an autonomous science? In: Toward a new philosophy of biology. Harvard University Press, Cambridge, MA, USA.

Shannon, C.E., Weaver, W. (1963). The mathematical theory of communication. University of Illinois Press, Champaign, IL, USA.

Tolkien, J.R.R. (1954). Lord of the rings. Allen & Unwin, Crows Nest, Australia.

14. The comprehensiveness of scientific models

Simplification versus completeness. Modeling helps us understand how nature works. To prevent a model becoming overly complicated, simplifying assumptions are made about what is included in the model. For example, a food-chain model may focus on 'species' and 'food'. The model is a map. It is not the territory. Can the comprehensiveness of the model-map be checked?

Nature's many faces. Imagine you are walking in a rainforest. High up in a tree some monkeys fight over a juicy fruit. The noise frightens a parrot that flies away. Many fruits have fallen on the floor, and have become moldy. A large millipede eats from the mold. A yellowish mushroom grows nearby. Its spores form a brownish blanket on the soil under its cap.

A model. Is it possible to model a complicated system like a rainforest? To answer this question it is practical to first elaborate the term 'model'. The generic definition I use is: a model is an entity representing another entity. This definition includes a wooden representation of a real sword children use in their play. This also includes the kind of computer simulation of an ecosystem I discuss here.

In research, the art of simulation is to construct the least complicated model that answers a specific research question in a trustworthy way. With respect to the word 'trustworthy', the goal of this SCIENCEBITE is to create a feeling of what is covered by a model, in comparison with what is not covered by a model. Such a comparison demands an overview of the entities and relationships involved in a model. Obviously, the myriads of interactions in nature make it hard to create a comprehensive overview. How can one tackle this problem?

The ecosystem as a system. A model of an ecosystem asks for a definition of the system concept. The Operator Theory defines a system as follows: a part of the universe (the 'object' of study) that is analyzed with a deliberate focus on objects and their relationships. I consider this focus a 'systemic perspective.' Fundamental objects in ecosystems that can be used in a model are operators of different complexity. Here one can focus on atoms and molecules, which are abiotic operators. One can also focus on organisms, which are biotic operators. If one uses the operator theory, the term organism can be specified as including the cells (bacteria/archaea), the endosymbiont cells (eukaryotes), the multicellulars based on cells, the multicellulars based on endosymbiont cells, and last but not least, the animals.

Specific interactions. Operators can interact in many ways. These interactions can be organized in a limited number of basic categories: interactions with abiotic entities, interactions with other organisms, interactions with offspring, and interactions with lumps of things (as was explained in SCIENCEBITE 1, the operator theory calls a lump of things a 'compound object').

When looked at from the interests of the organisms involved, interactions can range from synergistic to antagonistic. The kinds of interactions that are possible will depend on the kinds of operators involved. The possibilities for interactions between low-complexity abiotic operators, e.g. atoms and molecules, are relatively limited. For organisms the world is much more complicated. Even low complexity organisms, such as bacteria can move, consume materials, produce toxins or enzymes, attach to substrates, etc. They can interact with larger organisms, e.g. by living in the intestine, or as a disease, e.g. leprosy. Still more complicated organisms are multicellular plants. And animals with neural networks can interact on the basis of neural reflexes, or, in animals with sufficiently complex brains, on the basis of thoughts. The ability to think opens a door to very complicated interactions, because thinking facilitates the construction and use of tools and the development of behavioral strategies.

Towards a grouping. The ecosystem can be mapped as a graph that connects all abiotic and biotic operators through their interactions. As was discussed above, the diversity of the kinds of interactions between different operators is enormous. To simplify things, one option is to lump the entities into groups, for example according to classes in the operator hierarchy.

Another option is to lump interactions on the basis of general properties. What do I mean by this?

Properties. Imagine a bromeliad growing on a tree. The heart of the bromeliad contains water. Several mosquito larvae swim in the water. In these examples the construction of one organism is relevant for the next organism. The interactions share the property of 'construction'. By analogy, one can focus on a tiger walking in the forest. A branch cracks under her paws. The sound alarms a monkey that yells loudly to warn nearby relatives. This event is based on auditive signals that are perceived by organisms as 'information'. Two other properties I consider relevant in systems are 'energy' and 'displacement'. Energy offers a link with thermodynamics. And displacement covers the worlds of space and time.

Independence and overlap. The above focus allows one to identify four generic properties that describe 'independent' aspects of every interaction in the world. The advantage of identifying independent properties, that is, properties that focus on fundamentally different aspects of one and the same interaction, is that one can recognize them in any kind of interaction. For example, to model interactions that lead to a food chain, one can identify energy and construction as the main properties. The reason is that food contains energy (joules) in various forms, e.g. fat or sugar. This energy allows the consumer to fuel its body. Food also offers building materials, such as nutrients, amino acids, vitamins, etc. These materials allow the consumer to construct its body.

Construction as the basis. Because every object in nature has a material/ physical basis, the property of construction is always present. The other three properties can be linked to constructions. For example, Einstein discovered a fundamental relationship between matter and energy in the form of $E=MC^2$. Normally however, it is not the Einsteinian energy we talk about when referring to the energy that is captured in a specific kind of matter. The energy contained in matter is more like the heat that is freed when burning gasoline, or during an explosion of dynamite, or when digesting sugar or fat. Any relationship that causes an object to move through space, offers a link to dispersal. Likewise the construction of a material object offers a link with informational aspects. Starting from material existence, the other properties can be derived.

Dimensions. As a shorthand expression for 'class of independent properties' I also use 'dimension'. When I use the term dimension in this way, I do not aim at positioning entities along independent axes in a quantitative way. Instead the aim is to emphasize that one can look 'through' the ecosystem from different 'sides'. Each property represents a side; a dimension. Each dimension highlights a perspective that offers a new, independent focus on the relationships. In principle, one has only constructed a complete model of a system if the model accounts for all properties along each of the four dimensions. In general, however, a less complete model will have greater utility.

DICE. As a summary of the four dimensions I use the DICE anagram: Dispersal, Information, Construction, and Energy (Figure 14.1). If I have been correct in identifying a minimal number of fundamental properties, every interaction between two entities in nature has aspects that fit one or more of these dimensions. So far I have not found a reason for adding a fifth dimension, but this could change in the future. Just as a simple test, one can examine a berry of the mistletoe that looks attractive to a bird, and that tastes nice. These are informational aspects of the bird-berry relationship. When the bird tries to eat it, the berry sticks to its beak. This is a constructional relationship. Then the bird may eat it, and fly away. A day later the seed is released in the bird's droppings. A dispersal relationship.

DICE-completeness. DICE has been developed with the aim of inspiring conceptual analysis. DICE allows for a positive check of what is included in a model, and a negative check of what is not included. The positive check evaluates which dimensions a model focuses on. For example, a model of a food chain focuses predominantly on constructional aspects (nutrients, amino acids, etc.) and energy (calories). It also involves an informational aspect, namely, who eats whom. And to find 'prey', a herbivore or predator must displace itself, or make use of a moving environment (e.g. flowing

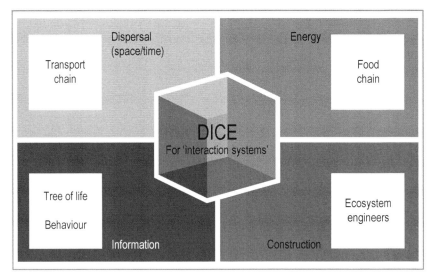

Figure 14.1. An example of how the four perspectives, or 'dimensions', of DICE (dispersal, information, construction and energy) can be applied during the analysis of an ecosystem. As the ecosystem consists of interacting operators, it is called an 'interaction system'.

water). The negative check focuses on what is not included. For example, a food chain model normally has a fixed and simplified list of the consumers and of those being consumed. Food chain models generally don't include evolutionary effects of genetic recombination and selection. What I see as the utility of using DICE is that it helps to raise awareness about the factors that are and are not included in a model. DICE also offers a heuristic for the identification of unexpected properties that play a role in relationships between organisms.

Further reading

Jagers op Akkerhuis, G.A.J.M. (2008). Analysing hierarchy in the organisation of biological and physical systems. Biological reviews 83: 1-12.

Jagers op Akkerhuis, G.A.J.M. (2010). The operator hierarchy. A chain of closures linking matter, life and artificial intelligence. Alterra scientific contributions 34. Wageningen Environmental Research, Wageningen, the Netherlands.

15. A swan song for the last black swan

After having observed thousands of white swans in ponds all over the world, a biologist may be tempted to conclude that all swans are white. However, the philosopher Karl Popper once said: a generalization like this cannot be trusted because someday a swan may be found that is black. Many laws in science are generalizations. Should one trust the scientist or the philosopher?

Black swan. If a limited number of observations serves as the basis for a general statement, this is called 'induction'. People use inductions every day, both in relation to objects (all swans are white) and in relation to events (water in a pond freezes when the temperature falls below zero Celsius).

Popper has become known for advocating caution with respect to inductions. He warned that irrespective of the number of confirming observations, the next observation can be a 'black swan' (Figure 15.1). When stated in this stringent way, large numbers of confirmations cannot verify an induction, while a single black swan leads to its falsification.

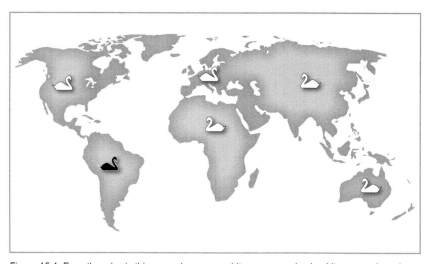

Figure 15.1. Even though – in this example – many white swans, and only white swans, have been discovered in the Eurasian continent, and North America, new discoveries may one day offer proof of the existence of one or more black swans in South America.

Based on the black swan logic Popper suggests that a theory is scientific if it allows falsifiable statements about reality, and advocates that the falsifiability of statements can be used as the demarcation criterion for distinguishing science from pseudoscience. But when an induction can always have a black swan that falsifies it, how can scientists and engineers confidently apply scientific laws? Or could it be that confidence is a matter of degree?

Induction. Before continuing the discussion of the black swan problem in science, a detailed picture is needed of the meaning of the word 'induction'. Generally speaking, an induction makes use of five criteria:

1. Two or more particulars (things or events). The particulars can, for example, be several large, white, waterfowl.
2. A category. The category can be 'swan'.
3. A general relationship. For example: 'swans are white'.
4. Independence of the category and the relationship. To prevent – partial – circular reasoning, the criteria for the category and the relationship must be independent. This means that for identifying a swan, one must not use whiteness, but other criteria, such as the shape and size, etc.
5. Identical conditions for the particulars and the inductions. To prevent one mistakenly classifying a dirty swan in a muddy pool as a black swan, identical conditions (*ceteris paribus*) are presumed. Accordingly, all swans must be observed under comparable circumstances, e.g. in a pond with clear water.

Before Popper, David Hume already emphasizes that an induction is an expectation, not a proof. Logically speaking the expectation that 'all swans are white' cannot preclude a black swan. An induction is based on premises that can at best offer a probability of a conclusion being true or false.

Deduction. In the context of black swans, it is relevant to when discussing induction, to also discuss its counterpart deduction. Deduction postulates that if one starts with true premises, and uses valid reasoning, the result is true. For example using the axioms of addition, of subtraction and the number 1, one can create all the numbers called integers using the following reasoning: 'Start with 1, and add or subtract one or more times the number 1'. In this way, one can use deductive reasoning to create the 'family of the integers' in a way that offers certainty about the properties of all the integers.

Another well-known deduction is the following: all people are mortal, Socrates is a human being, Socrates is mortal. As long as the premises are true, and the reasoning is valid, deduction leads to a true outcome. Assuming one can use deductive reasoning, there is no place for a black swan.

Premises. Induction and deduction represent logical approaches. Both approaches can be applied to theoretical situations and to situations in the real world. When being applied to the real world, the outcome of both induction and deduction depends on the quality of the data. The following examples demonstrate this point.

A famous story by Bertrand Russell is the following. Imagine a turkey that is fed every morning by a farmer. By induction, the turkey becomes more convinced each day that feeding happens every morning. On Christmas morning, however, the farmer acts differently, and the turkey has a fatal black swan experience. Here Russell's example ends. The turkey's focus on farmer and food is an example of Hume's analysis of empirical observations of the form: 'if A then B'. The turkey's information is limited to observations of a farmer who brings food every morning. To turn his induction into a more detailed prediction, the turkey would need more data.

If we assume that the turkey could have been informed in more detail about the behavior of the farmer at Christmas time, it could have *deduced* a probability of it being slaughtered. This would require, however, true information about the premises, and a valid rule. For example, the turkey would have to be sure that the farmer always – without exception – kills his turkeys on Christmas morning, and that the current day is the 25th of December, which day is Christmas. Through the use of additional information, the turkey's death is no longer a black swan, but a calculatable chance event, that can be viewed as a 'greyish swan'. Such a deduction, however, requires true information about the farmer's behavior, and the date.

The above real world examples of induction and deduction suggest that in both cases information is required about the quality of the data, and about the premises. As a cause-effect relationship allows one to connect events, and because practically all real world phenomena are the products of causal events, causal relationships offer valuable information for inductive and deductive predictions. The need for causal information leads to the question of how people observe the world and extract causal relationships from what they observe.

The personal perspective. During their observations of the world, both turkeys and people necessarily apply a personal perspective. The personal perspective is important for the quality of observations. The reason is that a person's knowledge depends on how he/she perceives things, how he/she ponders about the trustworthiness of observations, and how he/she integrates information into a mental model of relationships in the world. As Immanuel Kant said: no one can transcend the bounds of his/her own mind. Kant's statement applies also when knowledge about the world arrives to us by other means than empirical learning, e.g. through discussions, or via textbooks. Even then, observing and learning remain processes that are filtered by our perceptions and understanding.

When the role of personal experience is exaggerated, this can lead to the assumption that the world exists only in as far as it is perceived by, or even during, human observation. If you look at it, the world is there. If you close your eyes the world is gone. If you open your eyes the world re-appears. This is of course a rather human-centered viewpoint.

The 'world' perspective. Instead of making the world dependent on your personal perception, you can also assume that the world around you exists independently of whether you have your eyes open or closed. In this perspective it is not you who constructs the world through your perceptions. Instead, you, and your perceptions, are part of the world.

If you view yourself as part of the world, you are a realist. A realist views the universe as having formed as the result of natural processes, which for the larger part happen (and have happened) independently of the presence of man. Accordingly, lightning causes thunder irrespectively of anyone listening. And the existence of fossils offers proof of the existence of past organisms long before the arrival of human observers. Eventually, humans mix in.

According to a realist, one can do science only because entities exist that have lawful behaviors. This viewpoint is emphasized by realists such as Roy Bhaskar. If entities did not exist, or did not behave autonomously in accordance with their kind, scientific experiments would produce haphazard and nonsensical results. There would be no patterns, no law-like regularities.

Combining viewpoints. The personal perspective emphasizes that human knowledge depends on observations that are influenced by experiences and knowledge. The realist perspective says that without the independent physical existence of things there is no basis for knowledge. By adopting a realist worldview while accounting for the relativity of perceptions, a scientist can gain knowledge about natural processes, and use this as a foundation for inductive and deductive reasoning. A powerful tool for combining various kinds of knowledge about the world with inductive/deductive reasoning is a causal, dynamic model.

Basically such a model is a theoretical instrument for integrating knowledge about structure and change. For example, one can ask: how is a house built? To answer this question a modeler would first need a list of the parts of a house, such as windows, and stones. In turn, these parts may consist of still smaller parts, etc. With knowledge about the relevant parts, and with the help of a causal model, the process of building a house can be reconstructed from data about separately observed construction steps. At any moment we can observe that a person uses clay to bake stones. Or that a person combines stones and cement to make a wall, etc. By linking many small fragments of causal knowledge we can create a reconstructive model of the building of a house.

Reconstructing the universe. While searching for data for the development of reconstructive causal models, humans shine a light-beam of theory-laden observation at an increasing range of here-and-now phenomena. At bigger-than-human scales, observations have expanded to the amazing pictures of distant galaxies in the universe as are made by the Hubble telescope, and to measurements of the background radiation and gravitation waves. At smaller-than-human scales, scientists have 'descended' from objects of the size of animals and plants, to tiny objects like cells, molecules, atoms, hadrons, and fundamental particles, e.g. quarks and the Higgs particle.

Knowledge about signals from the edges of the universe in combination with observations of increasingly small particles allow scientists to fill in increasingly ancient and increasingly fundamental aspects of the retrospective causal model of the universe. With the help of such a model, all the current observations can be linked into a reconstructive 'chain' that allows us to speculate about past physical processes in times when humans did not yet exist, or future processes in times when humans may exist no more.

A swan song for the last black swan. In the current age, causal understanding rapidly expands to cover the entire range from fundamental particles to the universe. With time passing by, reconstructive theory of increasing detail and scope will become available. The space for an unexpected observation, an empirical black swan, is constantly decreasing.

Of course, scientists will always have to deal with inherent natural limits to scientific models, such as the limits that result from quantum uncertainty, from self-organized criticality, or from chaos theory. Within the boundaries of such inherent natural limits, human knowledge will in the future allow detailed causal explanations for any kind of real-world empirical observation. By improving their causal models, scientists will one day be able to offer bottom-up causal reconstructions of the entire history of the universe. I suggest, therefore, that scientific activity will ultimately allow scientists to compose a swan song for the last black swan.

Further reading

Benton, T., Craib, I. (2001). Philosophy of social science. The philosophical foundations of social thought. Chapter 4: Science, nature and society: some alternatives to empiricism. Palgrave, New York, NY, UK.

Bhaskar, R. (1977). A realist theory of science. Routledge, London, UK.

Derksen, A.A. (1980). Rationaliteit en wetenschap [Rationality and science]. Van Gorcum, Assen, the Netherlands.

Hume, D. (1975). Enquiries concerning human understanding and concerning the principles of morals (3rd Ed.). Oxford University Press, Oxford, UK.

Lakatos, I., Musgrave, A. (1974). Criticism and the growth of knowledge. Cambridge University Press, Cambridge, UK.

Popper, K. (1963). Conjectures and refutations: the growth of scientific knowledge. Routledge, London, UK.

Russell, B. (1902). The problems of philosophy. Holt & Co., New York, NY, USA.

https://scensci.wordpress.com/2012/11/20/poppers-falsification/

https://plato.stanford.edu/entries/induction-problem/

http://www.geo.uu.nl/fg/mkleinhans/teaching/geologica2eigner.pdf

16. Internal criteria for doing science

Relevancy. When is an activity 'scientific'? While defining science is a challenge, the answer is of practical importance, especially now that some people seem to think that science is 'just another opinion'. In such cases it is relevant to explain that scientists use norms for scientific activity while aiming at the production of an increasingly rich and insightful model of the world. Much work has been done already by philosophers and scientists to identify criteria for scientific activity. Therefore, my aim here is to offer an overview. What I emphasize is the value of differentiating between internal and external criteria.

Science is an activity. People 'do' science, hence science is an activity. Here I focus on scientific activity, in the larger context of a plurality of sciences where science is an umbrella term for all the different scientific disciplines, from physics to sociology, and for all the different aspects of the scientific endeavor, such as science as an 'enterprise', as a 'way of living', as a 'societal phenomenon', etc. The cartoon in Figure 16.1 offers an example of the plurality of science.

The roots of science. The word science stems from the Latin verb '*scire*', which means: 'to know'. In order to gain knowledge about entities and processes in the world, scientists are basically involved in two kinds of activities:

1. Scientists perform analyses and create models in relation to known territory. Our knowledge of known territory increases each time that observations confirm what we already knew, as well as when observations demand revision of our insights.
2. Scientists willfully explore unknown territory with two aims: (a) the disclosure of currently unknown properties of known phenomena; and (b) the disclosure of unknown phenomena. Unknown phenomena are mostly theoretically predicted and/or observationally suggested, e.g. 'black matter'.

Figure 16.1. Fokke and Sukke know what science is about (cover De bètacanon van Fokke & Sukke. J. Reid, B. Geleijnse and J.-M. van Tol, 2008. Reproduced with permission).

Almost all scientific observations depend strongly on the engineering of suitable equipment, and/or of new objects or new environments. Accordingly, theory and engineering are intertwined parts of science.

Some well confirmed views on science. Scientific activity depends on the world, and on our activity. A well tested view about the world is that a physical universe exists that contains entities that exhibit regular properties. If properties are never regular, scientific experiments can only produce haphazard results. Another well tested view is that entities exist and act regularly also when no human is observing them. Otherwise interactions that take place without human interference could not be explained. For example, if I close my eyes while walking, I can't see the world anymore, but I can still walk into a wall.

This shows that by not perceiving the world you will not make it disappear. The critical realist Roy Bhaskar points out that people sometimes overlook these premises for doing science (see also SCIENCEBITE 15).

Finally, to work efficiently, and to produce trustworthy results, scientists have developed criteria for scientific activity. Such criteria are not a gift of nature, but are the product of experience and philosophy. If one views science as a pluralistic concept and a non-essential category, the most practical and efficient way to advance discussions about what defines (a particular branch of) science is to work with a revisable list of criteria.

Internal and external criteria. Past debate has resulted in different listings of criteria for doing science (e.g. Isaac Newton, Herman De Regt and Hans Dooremalen). Here I organize these criteria in two groups: 'internal' and 'external'. Internal criteria form what I view as the minimum core of science. A researcher must respect internal criteria regardless of whether working alone or in cooperation. External criteria such as acceptance, credibility, patents, intersubjectivity, publications, etc., involve cooperative aspects, and are relevant in a social or economic context. I focus here on the internal criteria because these can be viewed as being the most inherently scientific.

Internal criteria for science. A scientist who focuses on internal criteria will in principle aim to respect all the criteria on a listing. I tentatively differentiate between 'interpretative' (logic, philosophy, etc.) and 'experimental' sciences (physics, biology, etc.). The following inventory may still be incomplete.

Internal criteria that are relevant for both interpretative and experimental sciences:
- Personal attitude. One should be critical and honest about one's practical and theoretical work.
- Generality. The importance of generality was already valued by Isaac Newton (1643-1727). A general theory (general not meaning 'vague') has greater potential to connect and integrate other theory. If a theory is not general, it is desirable that one indicates its scope.

- Efficiency of reasoning. Ockham's razor (William of Ockham, 1287-1347), as it is used today, advocates that in cases where two theories explain the same data with identical accuracy, the progress of scientific reasoning is easier, and more efficient, if one chooses the theory that has the least assumptions. This criterion leads to simplicity/parsimony/stinginess, and safeguards scientific theory from the accumulation of unnecessary assumptions and loads of alternative theories.
- Logical consistency. Scientists aim at constructing explanatory theory about events in the world. Theory is the logical model that connects observations. To preclude logical errors and category mistakes scientists strive towards logically consistent theory and a stringent use of categories.
- Terminology. To preclude vagueness, a clear and/or unambiguous terminology is sought for. The operator theory (see SCIENCEBITE 1) is relevant here, as it reduces ambiguity by introducing a fundamental distinction between the basic 'objects', the 'operators', e.g. atoms, molecules, cells, multicellular organisms, organisms with neural networks, and systems that consist of interacting operators without being an operator; the so called 'interaction systems', e.g. stones, cities, planets.
- Data. Care must be taken with respect to the use of data in a theory. One will have to take into account that data may result from 'indirect' observations and can be 'contaminated' with contextual information. All observations (including words in texts) are mediated by e.g. human senses, language, measuring equipment, the environment, personal and theory-based preconceptions, etc.

Additional criteria for sciences based on experiments and observations; the 'experimental' sciences:

- Mental groupings and classifications. Our thinking depends on mental models (representations, universals and categories, e.g. my dog, hunger, the color red, moral, etc.). Inside the neural network of the brain, every concept has a physical existence that can be studied through effects on physiology and activity, e.g. thinking, behavior, communication, etc. Some concepts, such as 'this organism' or 'my dog', have physical instantiations in the world. Other concepts do not exist physically in the world outside our head, e.g. 'love', 'Hobbit' and 'thousand'.

- Repeatability and verification. The logical positivists/empiricists (1920-1930) introduced an empirical cycle that includes: 1. A theory, 2. A theory-based prediction, 3. An experiment to test the prediction, 4. Comparison of experimental result with the prediction. If the outcome is true, and the experiment can be repeated, this adds credibility to a theory (which is not equal to proof).

- Falsifiability and improvement. Karl Popper (1972) argues that confirming observations cannot prove that a theory is correct, because any subsequent observation may yield a 'black swan' (however, see SCIENCEBITE 15). Because of this, one must focus on tests that – in principle – can debunk one's theory. Challenging and precise statements are preferred because these imply greater falsifiability. Later Imre Lakatos suggested that improvement – instead of debunking – is aimed at. As a rule, the theoretic improvement through reconceptualization of an idea that is central to a theory is a rare event, while peripheral ideas can change more rapidly.

- Causal explanations. Scientists use current observations (things and processes) as a basis for the causal reconstruction of past things and processes, and for causal predictions of future ones. In relation to this criterion, David Deutsch suggests that naturalistic scientists aim at maximizing the explanatory power of their models.

- Existence. Experimental scientists mainly focus on evidence for the physical *existence* of things. The reason is that it is generally hard to prove non-existence. Despite this, once it is positively identified by means of a scientific observation, any phenomenon will be added to the body of science, no matter how rare, local, temporally limited, hard to interact with, etc., it is (e.g. a neutrino). If no test has ever offered support for its physical existence, and/or if it has properties that defy general and fundamental natural principles, and/or if it adds nothing to these principles, science will consider a phenomenon to belong to the class of things that are hypothetical, supranatural, redundant, and/or not amenable to scientific investigation.

Internal versus external criteria. The above internal criteria, which are related to the quality of science as such, stand apart from the external criteria. The external criteria are directed at societal imbedding and impact. Examples of external criteria are: (1) the capacity to exchange ideas with others, the so-called 'intersubjectivity' of science; (2) the ability of a scientist to write grant applications that are positively evaluated; (3) the number of people citing a publication; (4) the ability of the scientist to present his/her work, and that of his peers, on radio and television; (5) the amount of money that can be made with a scientific discovery; (6) the fit of an idea to a currently dominant paradigm; (7) the patenting of an invention; and (8) the technological and societal implications, etc., etc.

Some 'external' criteria can be understood as the implicit results of the internal criteria. For example, one can view innovation as an external criterion in the sense that society expects scientists to produce innovations. However, innovation is already an implicit product of scientific activity. After all, the above criteria for science, notably honesty and improvement, will cause novel discoveries, and hence innovation.

Further reading

Bhaskar, R.A. (1997). A realist theory of science. Verso, London, UK.

De Regt, H., Dooremalen, H. (2008). Wat een onzin. Wetenschap en het paranormale. Boom, Amsterdam, the Netherlands.

Deutsch, D. (2014). Simple refutation of the 'Bayesian' philosophy of science. Available at: http://www.daviddeutsch.org.uk/blog/.

Lakatos, I., Musgrave, A. (1974). Criticism and the growth of knowledge. Cambridge University Press, Cambridge, UK.

Westfall, R.S. (1998). Never at rest. A biography of Isaac Newton. Cambridge University Press, Cambridge, UK.

17. Extending and generalizing evolution

Why extend the modern evolutionary synthesis? As Darwin had no access to detailed knowledge of genetics, he phrased his ideas about variation, selection and evolution in rather broad terms. This is why he spoke about the theory of 'descent with modification through variation and selection'.

When it comes to the generality of the concept of Darwinian evolution, however, Darwin's lack of detail can be viewed as an advantage. The reason is the following. Around 1940, the discovery of Mendelian inheritance and genetic mutations added detail to Darwin's theory. To include the new information, the theory of evolution was upgraded to what became known as the 'modern evolutionary synthesis'. This synthesis caused a narrowing of the focus towards genetics.

Currently, as the result of recent discoveries, including epigenetics and plasmatic inheritance, some evolutionary biologists feel the urge to broaden the perspective again, creating what is called an 'extended synthesis'. This begs the question of what kinds of 'extensions' are needed? Should the extensions remain limited to biology? Or should evolutionary theory be expanded to also include phenomena outside biology?

Extending evolution outside biology would allow evolutionary thinking to include things like thoughts, tools, pictures, computer viruses, etc. A general approach to evolution that extends beyond biology would only be possible, however, if the biological terminology in which Darwinian evolution is currently phrased were replaced by more general concepts. Can such an extended and generalized approach still be called Darwinian evolution? I think the answer is 'yes'.

What is Darwinian about Darwinian evolution? When speaking about Darwinian evolution, people refer to Charles Darwin's and Alfred Russel Wallace's efforts to provide a causal explanation for the existence of species. Species are reproductively isolated groups of (sexual) organisms that look similar when they are in the same phase of their life cycle. Key aspects of Darwinian evolution are:

1. Descent with modification. This implies that offspring are produced that are not identical to the parents.
2. Variation. This implies differences between parents and offspring and/ or between offspring.
3. Selection. This implies differential mortality of variable offspring when they have to survive in a natural or cultural environment.

Darwin defines selection as '*this preservation of favorable variations and the rejection of injurious variations*'. Preservation can be interpreted as survival, in particular survival until reproduction. Rejection can be interpreted as mortality, in particular mortality before reproduction. Variation of offspring in one or more properties, in combination with differences in survival until reproduction in association with these properties, allows one to establish the phenomenon of selection. After all, if one follows Darwin's description of selection, mortality of an individual organism is not selection, but just mortality.

On variation and selection. Both variation and selection can be analyzed from an internalist, an externalist, and an interactionalist point of view. Firstly, when variation is viewed as being caused 'from within', one can focus e.g. on mutations of the DNA, or the impact of ageing on the color of one's hair. Secondly, variation can be viewed as being caused 'from the outside', e.g. when low winter temperatures cause hares to grow a thick fur. Thirdly, the interactionalist viewpoint applies to a hare that has brown fur in summer and white fur in winter. The change in color can best be explained as the result of the interaction with predators that have little difficulty spotting a brown hare in a white snowy landscape.

Just like variation, one can view mortality before reproduction as the result of internal factors, such as genetic defects or a malfunctioning auto-immune response, as the result of external factors, such as an attack by a lion, and as an interactional property, for example when a hare that keeps its brown fur in winter gets eaten by a predator. A comprehensive theory of evolution in biology ideally has room for all these perspectives.

Extending what? Darwin's model makes use of organisms, descent, variation and selection. But before one can decide about whether or not such a model needs extending, or about how new aspects can best be added, the model itself needs to be described in a stringent way. It is my conviction that the construction of an unambiguous model of Darwinian evolution would profit from a focus on individuals, instead of on populations. And I advocate that it may be easier to describe the concept of evolution by means of a pedigree graph that meets specific criteria, than to define evolution as a process.

The reasons why I suggest the above two viewpoints are: (1) individuals (not populations) are the entities that are involved in reproduction; and (2) The terms variation and selection refer to a comparative assessments of the properties of organisms (not to a process).

Above I already indicated that Darwin's definition of selection was the preservation of favorable variations and the rejection of injurious variations. When described this way, selection involves an assessment of whether or not individuals that belong to one generation in a pedigree vary in their properties (such that they can be recognized as variations) and in achieving reproduction or not (preservation versus rejection). Both assessments focus on 'differences' between offspring. And both differences can only be assessed if one has observations proving that the offspring are not identical.

The criteria for selection thus involve the *assessment* of differences between offspring. Such an assessment is an observation. Neither the parents nor the offspring carry out selection. What the parents and offspring can do is to reproduce or die. Realizing that selection does not refer to a process but to an assessment, I started looking at Darwinian evolution as a special pedigree, namely one in which selection can be observed as a pattern. If Darwinian evolution is viewed this way, variation no longer needs to be mentioned separately, because selection presumes the presence of variation.

The above demonstrates that the criteria for Darwinian evolution refer both to processes (reproduction) and assessments (variation, selection). In relation to this insight I stopped writing in my work about 'the process of Darwinian evolution', and started using a 'Darwinian pattern of evolution' instead. In this view, the pattern defines selection, instead of the pattern being caused 'by' selection.

In my view, one advantage of defining Darwinian evolution as a pattern of descent with selection is the following: if one makes use of a pattern, all criteria for evolution can be linked to that pattern. This possibility creates an integrated conceptualization to which all criteria can be linked. If one defines Darwinian evolution as a pedigree with a special pattern of descent, one creates a solid basis for discussions about extension and generalization.

Using a smallest pattern as a foundation. From the perspective of how to make things easy, it is an advantage if an 'extension' of a theory can be founded on a 'least complicated, basic version' of the theory. From such a basic version, any and all extensions can be constructed. Viewing evolution this way may require some mental adjustment, because it implies that one replaces the idea of a single definition for 'the process of Darwinian evolution', by a family of patterns of descent, *some of which* classify as Darwinian patterns. This is a radical change in perspective, but, as I hope to demonstrate convincingly, one that is quite useful if one aims at a precise description of what is viewed as 'Darwinian evolution', and if one aims at the construction of an extended theory.

The smallest biological pattern. In a *smallest* Darwinian pedigree pattern, *all aspects must in principle be smallest*. And because it has a minimal structure, the smallest pattern is a practical basis for discussions about extensions. As an example of a smallest pattern of Darwinian evolution in biology, one can start with an imaginary example organism (Figure 17.1). When reproducing, this example organism can produce either one or two offspring. Variation is assumed to occur in the smallest form: only one of the offspring differs in a single aspect. The variation is not necessarily heritable. Impact of the organisms on the environment, and impact of the environment on survival is implicit. Selection is reduced to its smallest form: a difference in whether or not two differing offspring reproduce (when defined this way selection implies variation). What results is a smallest pattern of Darwinian evolution: a single parent produces two offspring which differ in at least one property, while only one of the offspring reproduces. Accordingly, the smallest mechanistic definition of Darwinian evolution is: 'A pattern of descent, with selection' (where selection is a pattern instead of a process).

Extensions of the smallest model. Starting with the smallest model, one can imagine a range of possible extensions. Instead of a bacterium, one may

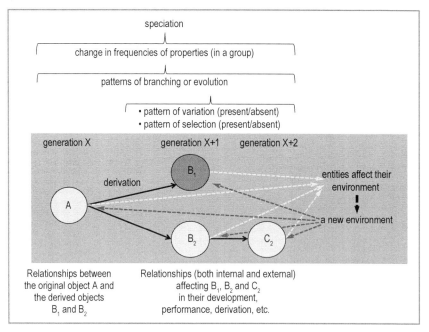

Figure 17.1. A smallest scale generalized model for Darwinian evolution. The example uses a generalized term for reproduction: 'derivation'. The derivation of entities, and a pattern of selection in interaction with the environment as affected by organisms, combine to a smallest scale pattern of Darwinian evolution. The smallest pattern can be linked to extensions and generalizations (circles = entities; black arrows = derivation; white dashed arrows = effects of entities on the environment; grey dashed arrows = effect of the environment on the entities; figure modified after Figure 16.3 in Jagers op Akkerhuis, 2016).

use more complex organisms, e.g. protozoa, plants, or animals. And instead of asexual reproduction, one can think of various forms of sexuality, and of kinds of multicellular fissure. While primitive organisms like bacteria have limited possibilities for development, and a limited life cycle, these aspects can be extended. Instead of a single variation, a broad range of potential internalist, externalist, and interactionist processes can be imagined as co-occurring causes due to which organisms vary in many different ways. Instead of a property being heritable or not, properties may be 'transferable' to different degrees. The effects of organisms on the environment, and the impact of the environment on the organisms, can be made explicit. One can include horizontal transfer of genetic and other information. And so on, and so forth. What remains constant in all these extensions is that any pedigree that classifies as Darwinian evolution must minimally comply with all the criteria of the smallest model.

A general terminology based on 'derivation'. Until now, the analysis has focused primarily on biology. However, a whole range of extensions can be thought of that would make patterns of Darwinian evolution applicable in the non-biological sciences. To accommodate for other entities than organisms, the major change that is needed is a generalization of the terminology. For example, instead of organisms one can focus on 'operators' or on 'compound objects'. The inclusion of 'interaction systems' (as defined in the operator theory) will not easily fit to the pedigree graph, but maybe special cases can be found that fit the logic.

Another term that lends itself to being generalized is reproduction. In Figure 17.1 I introduced the broader term 'derivation'. Derivation stands for all the mechanistic ways in which a next-generation entity can be produced 'after the image of' a parental entity. Interestingly, when using derivation, it becomes necessary to discuss in what way, and to what extent, a derived entity looks like its original. This is a topic that is rarely discussed in biology, because the reproduction of organisms sets limits to the forms of the offspring. Yet, in complex life cycles the subject is relevant, because multicellular organisms frequently have unicellular offspring, e.g. egg and semen.

Using derivation, Darwinian evolution can also incorporate pedigrees in which the next-level entities are produced by other entities, in a process that is known as scaffolded production. The scaffolded way of derivation applies to e.g. computer viruses, tools, drawings, etc. Derivation also comprises the combination of objects, called 'combinatorial evolution' by William Brian Arthur. Using the new terminology, descent can be viewed as the result of derivation. The smallest generalized definition of the special kind of evolution that is known as Darwinian evolution now becomes: '*a pattern of derivation, with selection*'.

In conclusion. When discussing extensions of Darwinian evolution, it helps to separately focus on the identification of a smallest possible pattern of

Darwinian evolution, on linking this pattern to more complicated patterns in a process of extension, and on generalizing the concept of Darwinian evolution by generalizing the ontology from organisms and reproduction to any entities that fit a pedigree combining the patterns of derivation and selection.

Further reading

Bickhard, M.H., Campbell, D.T. (2003). Variations in variation and selection: the ubiquity of the variation-and-selective-retention ratchet in emergent organizational complexity. Foundations of Science 8: 215-282.

Brian Arthur, W. (2009). The nature of technology. What it is and how it evolves. Penguin, London, UK.

Darwin, C.R. (1859). On the origin of species by means of natural selection, or the preservation of favoured races in the struggle for life (1st Ed.). John Murray, London, UK.

Huxley, J.S. (1942). Evolution: the modern synthesis. Harper, New York, NY, USA.

Jablonka, E., Lamb, M.J. (2005). Evolution in four dimensions: genetic, epigenetic, behavioral, and symbolic variation in the history of life. MIT Press, Cambridge, MA, USA.

Jagers op Akkerhuis, G.A.J.M. (ed.) (2016). Evolution and transitions in complexity. The science of hierarchical organization in nature. Springer, Cham, Switzerland.

Laland, K.N., Uller, T., Feldman, M.W., Sterelny, K., Müller, G.B., Moczek, A., Jablonka, E., Odling-Smee, J. (2015). The extended evolutionary synthesis: its structure, assumptions and predictions. Proceedings of the Royal Society B 282 (1813): 20151019.

Lewontin, R.C. (1970). The units of selection. Annual Review of Ecology and Systematics 1: 1-18.

Losos, J.B. (2013). What is evolution? In: J.B. Losos, D.A. Baum, D.J. Futuyma, H.E. Hoekstra, R.E. Lenski, A.J. Moore, C.L. Peichel, D. Schluter and M.C. Whitlock (eds.). The Princeton guide to evolution. Princeton University Press, Princeton, NJ, USA.

Pigliucci, M. (2007). Do we need an extended evolutionary synthesis? Evolution 61:2743-2749.

Stearns, S.C. (2013). Natural selection, adaptation, and fitness: overview. In: J.B. Losos, D.A. Baum, D.J. Futuyma, H.E. Hoekstra, R.E. Lenski, A.J. Moore, C.L. Peichel, D. Schluter and M.C. Whitlock (eds.). The Princeton guide to evolution. Princeton University Press, Princeton, NJ, USA.

Wallace, A.R. (1870). Contributions to the theory of natural selection. Macmillan and Company, London, UK.

18. Organism versus holobiont

Holobiont. Biologists and ecologists generally view organisms as the basic units of biological organization and evolution. A recent viewpoint now suggests a theoretical innovation: researchers no longer use the organism as the basic unit of biology, but redefine their thinking in terms of the organism and all the bacteria and fungi that, as symbionts, mutualists, commensalists or parasites, live on the skin, in the intestines, or in its cells (e.g. mitochondria and/or chloroplasts) (Figure 18.1). The new unit is called a 'holobiont'. The term holobiont stems from the work of Lynn Margulis on symbioses and combines the concepts of holos, meaning 'whole', and biont, meaning a living entity, or a 'unit of life'.

On the one hand, one can welcome the holobiont as a new and interesting holistic concept that may leverage creativity and innovation. On the other hand, the introduction of the term holobiont would imply a new categorization in biology that does not overlap with the organism concept. When suggesting a new categorization one must analyze relationships with existing theory about what have so far been viewed as 'units of life', and with

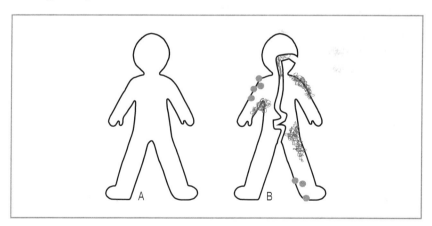

Figure 18.1. Organismal versus holobiontic perspective. A: the multicellular animal as a 'clean' entity. B: the holobiontic version of the animal, with different kinds of organisms living on its skin and/or in its gastrointestinal tract, or living in each cell of the organism.

established principles such as selection and evolution. Here I focus on such considerations.

Holobiontic thinking. Several recent observations have inspired researchers to consider thinking in terms of holobionts. A paradigmatic discovery was made during the study of corals. The polyps of coral depend for their metabolism on flagellate unicellular algae of the genus *Symbiodinium* that live inside coral cells. These algae are facultative inhabitants that are acquired from the sea water, and that can be expelled from the polyp, resulting in a 'bleached' appearance. Without its symbiont the polyp is short of energy and will eventually starve to death. Corals have a long life-cycle and evolve slowly. However, the composition of the algal symbionts can adapt to altered circumstances in less than a generation, suggesting that viewing a coral as a holobiont is relevant in the context of selection, and therefore for evolution. Host-microbiota relationships relevant for selection are also observed in other cases, e.g. gut bacteria in mammals, light-producing *Vibrio* bacteria in squid, and beneficial *Regiella*, *Hamiltonella*, and *Serratia* bacteria in pea aphids.

The intertwined functioning of the host and endosymbionts makes the ensemble appear as a single unit of selection; hence the holobiont idea. One can now ask: How strong must a symbiotic relationship be to consider the holobiont as a unit of selection? And what is the difference with considering it as a unit of evolution? To answer such questions insight is needed into the relationships between the terms biont, selection and evolution.

Biont versus organism. As Margulis indicated, the term 'biont' refers to a 'unit of life'. What is a unit of life? Is a spermatozoid a unit of life? Is a bacterium a unit of life? Is a cow, with all its intestinal microbiota, a unit of life? Is a tree in the rainforest, with all the attached epiphytes, and insects, mosses and fungi living in or on the epiphytes, a unit of life? Is a protozoa with its mitochondrial endosymbionts a unit of life? Is an entire forest a unit of life? In my view the answering of questions like these can profit from criteria that help distinguish between single individual organisms and groups of organisms. I discuss some examples that clarify this difference.

Organism versus group: a bacterium. Single organisms exist in different forms. The smallest single organisms are the bacteria and archaea. These are single cells, and single units of life.

Organism versus group: a protozoa. Examples of unicellular organisms with a more complicated structure than bacteria, are protozoa. A protozoan host harbors one or more unicellular endosymbionts, e.g. algae and/ or mitochondria. Defining the organism concept in the case of protozoa requires a choice. One can choose to view a protozoa as a single organism or as a symbiotic group of organisms. To decide between organism or group, it helps to distinguish between facultative or obligatory symbioses. A *facultative symbiont* can be acquired from the environment and can at some later moment be actively exported from the protozoa. The import and export allows for horizontal transmission of facultative symbionts between host cells. While a life without a host or without endosymbionts may reduce longevity, the host and the symbionts can in principle part and live separately, and find other endosymbionts or other hosts. As a result the host and symbionts can form independent evolutionary lineages. On these grounds, the situation can best be viewed as an organization that involves two or more organisms: the host and the facultative endosymbionts. Things differ in *obligate symbionts*. These cannot leave the cell, and, as a consequence, horizontal transmission does not occur independently of vertical transmission. Reproduction of the host represents a 'channel' for the endosymbionts towards the offspring. As a rule, the endosymbionts can only form evolutionary lineages 'inside' this channel. Accordingly, the evolutionary lineage of the symbionts has become trapped inside the lineage of the host cell. Now the endosymbionts are no longer free to leave the cell. They can be viewed as having become a special kind of (evolving) organelles and as parts of their host. Because of obligate structural and evolutionary dependency, a protozoa with endosymbionts can be viewed as a single organism.

Organism versus group: a multicellular organism. As with protozoa, a definition of whether a group of cells is a single organism requires a choice. To facilitate this choice I distinguish between 'pluricellular' and 'multicellular' groups of cells.

In a *pluricellular group* of cells the unity is limited to attachment between cells. While the cells will generally be of the same species, a clonal descent is not required. A pluricellular group may form when cells that previously lived independently aggregate. An example is the slug of a slime mold. While a pluricellular group can be selected as a group, for example when the entire group is eaten by another organism, the group is not necessarily a unit of evolution. The reason why this is not the case is that each cell that enters the group carries its own genes, and contributes to its own evolutionary pedigree. A pluricellular organization therefore represents a group of interacting organisms.

Things are different in a *multicellular group* of cells where cells are clonal offspring of a common parental cell, and form a single unit in which the cells are connected though plasma connections. For this reason, a multicellular group can be viewed as a single unit of selection. Reproduction may involve both single cells and/or small multicellular lumps. When the criteria of plasma connections between the cells, a common outer membrane, and mutual dependence of these two aspects are met, this implies dual closure (see SCIENCEBITES 1 and 9). When dual closure is present the cells of a multicellular organization can be viewed as a single organism.

Holobiont: organism or group? The above discussions of protozoa and multicellular organisms offer criteria for distinguishing between individual organisms and groups of organisms. To classify as an organism the contributing cells must form a physical unit, either as a host with endosymbionts, or as clonal cells connected through plasma strands, and the connections between the cells must be obligatory. If these criteria are met, the evolutionary lineages of the organism and its constituting cells are united during vertical transmission, and cells of the multicellular organism cannot at the same time take part in horizontal transmission.

These criteria can also be used for determining whether a holobiont is a 'unit of life' or not. But before we can step forward to a decision, it is relevant to consider the gradient that may exist in nature as to how strongly cells depend on each other, and to realize that over many generations such dependency may slowly become strengthened. The assumption behind using the above criteria for the identification of multi-cell organisms, such as protozoa and multicellulars, is that the relationship between the contributing cells, however gradual the evolution of their dependency, can at any moment be evaluated as being either present or absent. I prefer an approach like this over other criteria such as physical attachment and/or reproduction as a unity, because those criteria leave more space for exceptions and vagueness.

In conclusion. If we now use the absence of horizontal transmission, and obligatory dependency of the cells resulting in an overlapping channeling of the evolution of the contributing cells, one can conclude that a protozoa represents a single organism because of the obligatory interdependency of the host cell and its endosymbionts. At the same time, a protozoa can be viewed as a holobiont with respect to bacteria that attach to its outer surface. Likewise, a multicellular organism represents a single organism because of plasma connections between the cells, including mitochondria or chloroplasts in these cells. At the same time, a multicellular organism can be viewed as a holobiont, because of the microbiota living on its body surfaces (e.g. skin, urinal tract, gut, intestines, nasal cavity, etc.).

The possibility of horizontal transfer of the microbionts that live in association with any holobiont demonstrates that there is no full overlap of the concepts of holobiont and unit of evolution. A holobiont combines different organisms, each taking part in its proper evolutionary pedigree. Because of this, the holobiont concept can be viewed as a term that amalgamates the organism as a unit of selection, with the organism as a patch of environment that other organisms live in or on. Analogous reasoning applies to the genes of the holobiont, called the hologenome, when compared to the genome of an organism.

Further reading

Baker, A.C. (2003). Flexibility and specificity in coral-algal symbiosis: diversity, ecology, and biogeography of *Symbiodinium*. Annual Review of Ecology, Evolution, and Systematics 34: 661-689.

Haynes, S., Darby, A.C., Daniell, T.J., Webster, G., van Veen, F.J.F., Godfray, H.C.J., Prosser, J.I., Douglas, A.E. (2003). The diversity of bacteria associated with natural aphid populations. Applied and Environmental Microbiology 69: 7216-7223.

Jagers op Akkerhuis, G.A.J.M. (ed.) (2016). Evolution and transitions in complexity. The science of hierarchical organization in nature. Springer, Cham, Switzerland.

Margulis, L. (1991). Symbiosis as a source of evolutionary innovation. MIT Press, Cambridge, MA, US.

Moran, N.A., Sloan, D.B. (2015). The hologenome concept: helpful or hollow?. PLoS Biology. 13: e1002311.

19. Organisms versus superorganisms

Organization above the organism. Beehives and termite mounds are fascinating structures. During their construction, many bees or ants work together in close cooperation. Groups of bees or ants called castes, carry out specialized tasks. Such cooperation and division of labor calls to mind the manner in which a multicellular organism depends on the cooperation of its many different kinds of specialized cells. Possibly because of this analogy, ecologists sometimes refer to a beehive or termite mound as a super-organism, or supra-organism. Here I discuss the advantages and disadvantages of such a perspective.

From organism to superorganism. I start by discussing the super/supra-organism concept. Both terms 'super' and 'supra' grammatically suggest that a colony of bees and ants is an organism, although of a 'super' or 'supra' kind (Figure 19.1). Interestingly, the classical terms 'colony', 'eusocial colony', or 'colonial organization' did not refer to an organism, but to a colony. This begs the question: why did some scientists start using the superorganism terminology?

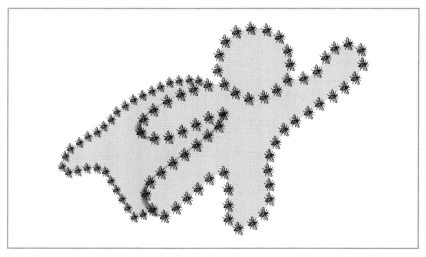

Figure 19.1. A group of cooperating bees is viewed in this picture as creating a super-organism.

A lack of unanimity. There is little unanimity about how to define the term organism. This may be the reason why it is used in different ways. To resolve the continuing discussions about how to define the organism concept, it has been suggested by e.g. Peter Godfrey-Smith that scientists must stop looking for a definition, and instead should arrange entities on a gradient from a lower to a higher degree of 'organismality'. The idea is that the more criteria of 'organismality' an entity meets, the closer its organization is to being an organism. What I see as a problem is that phenomena similar to organismal properties are common, both in living and non-living nature, and that as a result, many things can be viewed as being organismal to some degree or in some way. This can cause blurred reasoning. Moreover, in any philosophy like this, the term 'organismality' still depends on the term 'organism', which term remains undefined. It looks as if the can is kicked down the alley without scoring.

Multicellular organisms. As long as the organism concept lacks a definition, any definition of an organism can be disputed, including the 'multicellular organism'. After all, if one compares the specialized cells in a multicellular organism with the bees of specialized castes in a hive, the analogy suggests that even a multicellular organism can itself be viewed as a 'superorganism'. However, if one applies this reasoning, a plant and a human being are an organism and a superorganism at the same time.

I use the examples of plants and humans as a demonstration of how unclear definitions can cause logical conflict. When terms are used in ways that are considered interesting or useful, without assuring logical consistency, the resulting state of affairs conflicts with the list of fundamental criteria for science discussed in SCIENCEBITE 16.

Organisms of differing complexity. Different proposals have been made to arrive at a definition that offers necessary and sufficient criteria for the term 'organism'. The term 'necessary' implies that whenever a thing is an organism it complies with the criterion. A criterion such as 'activity' is not a necessary criterion, because even though a frozen bacterium is inactive it classifies by all structural means as an organism. After all, a frozen bacterium still has the potential for becoming active again. This potential can be tested by thawing the bacterium. The other term, 'sufficient', implies that the criteria used always allow an unambiguous conclusion. For example, the requirement of having a 'metabolism', when viewing metabolism as the combined chemical processes that have been tuned by evolution in the direction of sustenance of the organization of an organism, seemingly fits any and all physiologically active organisms. Therefore it seems a sufficient criterion. However, a frozen bacterium debunks this reasoning, because its icy state arrests any form of metabolic activity.

Organism definition of the operator theory. The operator theory can help to define the organism concept in a necessary and sufficient way. Basically, and from the bottom up, every operator in the hierarchy is defined by a new 'dual closure'. Dual closure demands compliance with three criteria: 1. A functional closure, 2. A structural closure, and 3. An obligate interaction between these two closures. Dual closure was discussed e.g. in SCIENCEBITES 1, 9 and 13. By definition, every operator has dual closure. And no other system can have dual closure, because if a system has dual closure, it classifies as an operator!

Dual closure thus is a necessary and sufficient criterion for all operators. This means that if one uses the operator concept as the basis for defining the organism concept, such a definition inherits the necessity and sufficiency of the operators. Based on this, one can define as organisms all the operators that are at least as complex as the cell, which are: the cell, the multicellular, the endosymbiont cell (classically termed 'eukaryote'), the endosymbiont multicellular, and the neural network organism. In this way, the long-standing problem of defining organisms of differing complexity can be resolved. Now that the organism concept was defined in an unambiguous way, we can return to the superorganism concept.

Organism or superorganism? In the following paragraphs I analyze three practical examples where I think it is relevant to distinguish between the (multicellular) organism concept and the superorganism concept.

Is a plant a superorganism? From the superorganism perspective, it has been suggested that any composition of cells represents a superorganism. This would imply that the organism concept *per se* only applies to single bacterial cells, protozoa already being super-organisms, while plants can be viewed as super-super-organisms. However, from the perspective of the operator theory, any operator from the level of the cell and up, is an organism, while the highest-level dual closure determines the level and the kind of its organization as an organism. From this perspective, a plant is an organism that has organs, such as a stem, leaves, and flowers, and these organs have parts, respectively the cells, and these cells have parts that take the shape of endosymbiotic mitochondria and chloroplasts, etc. According to the operator theory a plant is an organism, not a colony. As has been explained before, this is because the cells of the plant are too tightly connected to be called a colony. It is because of the dual closure based on bi-directional plasma strands between the cells in combination with a shared membrane, that the cells of a plant are no longer viewed as 'just' a colony of attached cells.

Is a social colony a superorganism? There exist many analogies between the organization of an organism and the organization of colonies of bees or ants. It is tempting, therefore, to compare an ant hill or beehive with an organism. However, according to the operator theory the existence of analogies does not warrant saying that a colony actually *is* an organism. That the operator theory does not view a colony as an organism is because every single ant or bee has dual closure at the level of a 'neural network organism' (or 'memon'), and accordingly already classifies as an organism. Meanwhile, the colony of ants or bees does not have dual closure, and for this reason it does not classify as an organism. However strongly integrated, by lack of dual closure a colony of ants or bees is just that: a colony. Moreover, when observing a colony of ants or bees, one can deduce that its organization is not structurally closed by means of an outer membrane that contains all the ants or bees. The organization of a colony results from interactions between individually countable, and individually behaving organisms. The colony is a colony. Adding the prefix 'super', and speaking about a colony as a superorganism, as if it were a special kind of organism, still doesn't turn a colony into an organism.

Is a city a superorganism? It will be clear by now that the operator theory applies the criterion of dual closure again and again. The simple question the operator theory asks when talking about a city, is whether or not a city has dual closure. The answer is negative. Accordingly a city is not an organism, and cannot be called a superorganism for that reason.

In conclusion. The above reasoning shows that there is a scientific need for the development of an ontology for different kinds of integrated organizations, e.g. beehives, cities, companies, etc. It would be practical if the words in such ontology would not suggest analogies with the organism concept. This is an interesting task for ontologists.

Further reading

Godfrey-Smith, P. (2010). Darwinian populations and natural selection. Oxford University Press, Oxford, UK.

Jagers op Akkerhuis, G.A.J.M. (ed.) (2016). Evolution and transitions in complexity. The science of hierarchical organization in nature. Springer, Cham, Switzerland.

Metz, J.A.J. (2013). On the concept of individual in ecology and evolution. Journal of Mathematical Biology 66: 635-647.

Moritz, R.F.A., Southwick, E.E. (1992). Bees as superorganisms: an evolutionary reality. Springer, Berlin, Germany.

Queller, D.C., Strassmann, J.E. (2009). Beyond society: the evolution of organismality. Philosophical Transactions of the Royal Society B 364: 3143-3155.

Seeley, T.D. (1989). The honey bee colony as a superorganism. American Scientist 77: 546-553.

Stebbins, G.L. (1969). The basis of progressive evolution. University of North Carolina Press, Chapel Hill, NC, USA.

20. Self-organization and the principle of least action

The principle of least action. In one of his famous lectures Richard Feynman says: '*When I was in high school, my physics teacher...called me down one day after physics class and said, "You look bored; I want to tell you something interesting." Then he told me something which I found absolutely fascinating,...The subject is this – the principle of least action.*' Why is the 'principle of least action' so fascinating? What is its relevance in system science and biology?

What is fascinating. Imagine you throw a stone. After the stone has left your hand, it follows a curved trajectory that ends somewhere on the soil. The path the stone follows to its 'destination' appears to be the shortest and the quickest of all possible paths. This shortest path appears to have the lowest value of a special quantity called 'action'. What is fascinating about the path of the stone, is that it shows that nature seems to have a preference for efficiency. And if this preference is a general rule, one may expect it to be present everywhere. When a leave falls from a tree, does it follow the most efficient path? And how about a river that meanders to the sea? Would there be a general efficiency principle, a principle of least action, that governs the movement of stones, leaves and water alike? Would the same principle also apply to organisms and their behavior?

The meaning of action. In physics, the term 'action' is the difference between kinetic energy and potential energy. This difference is summed over time (physicist call this an 'integral'). It is easy to explain this, if one uses the example of a ball that rolls down a hill (Figure 20.1). At the onset, the ball lies at the top of the hill. It does not move, thus its motion-energy, its 'kinetic' energy, is zero. While still at the top, the height-energy, its 'potential' energy, is at its maximum. When the ball starts rolling downhill, the loss of height implies a reduction of its potential energy, while the gain of speed causes an increase of its kinetic energy. As total energy is always conserved, the sum of kinetic and potential energy remains constant.

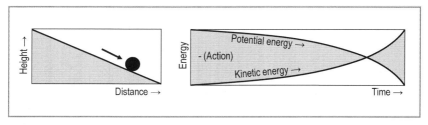

Figure 20.1. Left: a ball rolling down a hill. Right: the time-dependent changes in the kinetic energy, and the potential energy of the ball, when it rolls down the hill. The surface in between the two curves is a measure for the overall action when the ball has reached the bottom of the hill.

A graph of the two energies of the rolling ball offers a way to depict action graphically (for graphical reasons I invert the sign, focusing on potential energy minus kinetic energy). In the graph, the distance between the curves reflects the action at any moment, while the surface between the curves reflects the summation over time; the overall action. The smaller this surface, the smaller the action.

Shorter path; less action. In Figure 20.1 the ball rolls down a hill with a 30-degree slope. The path is long, the acceleration slow, and the action is high. If we choose a steeper hill, the path becomes shorter and the ball accelerates faster (Figure 20.2, example 2, with 45-degree slope). As a consequence, the action is less. And if the ball would simply fall down a vertical slope, the path would be shortest, and the acceleration would be at its max (Figure 20.2, example 3). The right part of Figure 20.2 demonstrates that the vertical path has the least action. If there is no obstruction, no 'hill', a ball falls towards the earth in the shortest time along the shortest path. This is the least-action path. What happens if in nature the least-action path is not immediately available because of obstructions?

Constraints. There is an analogy between the example of the ball rolling down a hill, and rainwater running down a mountain slope. The shortest path for rainwater to follow is a straight fall from the clouds, through the air, into the sea. However, if rain falls on a mountain, the water must follow a longer path because soil, rocks and other things are in the way. While flowing horizontally, the water will erode the soil, forming gullies, streams, and rivers.

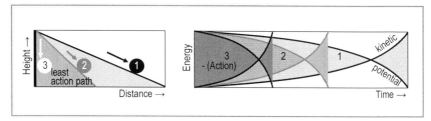

Figure 20.2. Left: three examples of the actual path of a ball rolling down a hill. The paths of the balls 1, 2 and 3 are in order of diminishing action. Right: the three surfaces indicate the temporal development of the kinetic and potential energy. There is a marked decrease in action when going from path 1 to path 3.

The Grand Canyon is a product of this process. The steep sides of the canyon allow the rainwater to flush down in a short period and over a short distance, and hence with little action. Through its wild flow the river at the bottom continues to grind deeper and deeper into the soil, bringing the river closer to sea level, and reducing the action further still.

Least action and self-organization. One can view the Grand Canyon as a result of 'self-organization'. The least-action path can be viewed as an attractor, pulling the river further and further down into the canyon (Figure 20.3). Meanwhile, many factors interfere with the outcome of the self-organization process, e.g. the irregular bulk movement of the flowing water, and differences in resistance to erosion of soil or rock. Some conditions will cause a river to meander. Other conditions will cause a river to branch into a complex delta. In both cases development towards least action dynamics in combination with constraints by the environment will determine the final pattern that will form.

The constructal law. A well-known approach to self-organization is the 'constructal law' of Adrian Bejan. This law focuses on systems in which a kind of medium, e.g. water, blood, air, etc., flows from one end to another. Such systems are called 'flow systems'. The currents in flow systems result from an external factor that imposes the flow, e.g. when rain and snow 'feed' a river. The constructal law is phrased as follows: '*For a finite-size system to persist in time (to live), it must evolve in such a way that it provides easier access*

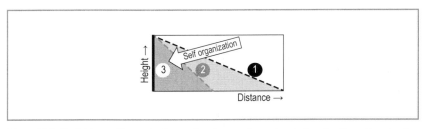

Figure 20.3. Explaining self-organization as the change of a system in the direction of least-action dynamics.

to the imposed (global) currents that flow through it.' Change of the system is towards: 'easier access to the imposed (global) currents'. If in the end the access provided were 100%, a current would pass through the system in the shortest possible time. At that moment the flow would follow the path of least action. In reality, systems are full of limiting conditions, and the flow can only reach a path of lowest attainable action. When a flow is constrained by limiting environmental conditions, like a river catchment directing the flow of a river, or when water must spread through veins to the entire surface of a leaf to maximize evaporation, or when air flows through tracheae in lungs, the lowest attainable action path will generally be fractal in shape.

In conclusion. While the principle of least action by itself does not predict what kind of patterns will emerge through self-organization, the principle helps to decide about the direction of change in the system.

Further reading

Annila, A. (2017). Evolution of the universe by the principle of least action. Physics Essays 30: 248-254.

Bejan, A. (1997). Advanced engineering thermodynamics. Wiley, New York, NY, USA.

Bejan, A., Zane, J.P. (2012). Design in nature. Doubleday, Toronto, Canada.

Brooks, D.R., Wiley, E.O. (1986). Evolution as entropy. Toward a unified theory of biology. University of Chicago Press, Chicago, IL, USA.

Chatterjee, A. (2015). Thermodynamics of action and organization in a system. Complexity 21: 307-317.

Feynman, R. (1964). The principle of least action. The Feynman lectures on physics 19. Available at: http://www.feynmanlectures.caltech.edu/II_19.html#footnote_source_1.

Georgiev, G., Georgiev, I. (2002). The least action and the metric of an organized system. Open Systems and Information Dynamics 9: 371-380.

Kauffman, S.A. (1993). The origins of order: self-organization and selection in evolution. Oxford University Press, Oxford, UK.

Kay, J.J. (1989). A thermodynamic perspective of the self-organization of living systems. In: P.W.J. Ledington (ed.). Proceedings of the 33rd Annual Meeting of the International Society for the System Sciences, Edinburgh, UK, Vol. 3, pp. 24-30.

Lambert, F.L. (1999). Shuffled cards, messy desks, and disorderly dorm rooms – examples of entropy increase? Nonsense! The Journal of Chemical Education 76: 1385-1387.

Lotka, A.J. (1945). The law of evolution as a maximal principle. Human Biology 17: 167-194.

Noether, E. (1918). Invariante Variationsprobleme. Nachrichten von der Gesellschaft der Wissenschaften zu Göttingen, Mathematisch-Physikalische Klasse 1918: 235-257.

Swenson, R. (1989). Emergent attractors and the law of maximum entropy production: foundations to a theory of general evolution. Systems research 6: 187-197.

21. Limits to scientific knowledge

How much do we (not) know? I often hear scientists say: *'There is still so much we don't know'*, or *'The more we know in science, the larger the domain of conscious ignorance'.* Interestingly, such statements beg the question of what we already know about our world, and how one can figure out how many things await discovery in the future. Is there a limit to scientific knowledge? (Figure 21.1). In this SCIENCEBITE I explore an approach that aims at handling questions like these in a structured way.

A mapping of realms of things. To deal with the question of what is known or not known to science, one needs a theoretical framework allowing one to reason about this question in a structured way. The framework suggested here is the operator theory. This theory assumes that most of what we know about our universe can be partitioned into three major fields. The

Figure 21.1. Is there a limit to scientific knowledge?

first field is that of special building-block systems, named 'operators'. Using the fundamental particles as a foundation, the range of the operators extends to neural network organisms. The second field is that of the 'inside' of every operator, its 'internal organization'. This second field covers, for example, all there is to know about the construction and functioning of an individual atom or organism. The third field focuses on the knowledge that can be gained about things that are constructed from interacting operators. Here one can think of the universe, a planet, and a car.

I am aware that e.g. the Big Bang, and empty space lay outside the above three fields. Furthermore, scientists may in the future discover systems that lay outside the reach of the operator theory. Such possibilities represent a known unknown. While acknowledging this particular uncertainty, I will use the operator theory as a relatively comprehensive, but not necessarily complete basis for discussing what can be known.

The operators. In my theoretic work I refer to the many different kinds of building blocks of the universe as 'operators'. Low complexity operators form the basis for more complex operators (see SCIENCEBITES 1, 8 and 10). Repeating this logic one can build a hierarchy of operators of different kinds, from fundamental particles, to hadrons, to atoms, molecules, cells, endosymbiont cells (the 'eukaryotes'), multicellular organisms of cells, multicellular organisms of endosymbiont cells, and, finally, multicellular organisms that are constructed from endosymbiont cells and that have a neural system. Every operator of a next kind has a unique, new combination of organizational properties that result from dual closure (e.g. SCIENCEBITE 13). The kind of dual closure defines the position of every operator in the operator hierarchy.

Being a member of the operator hierarchy sets the operators apart from all other systems in the universe. As far as I have been able to check, the hierarchy of the operators offers a complete classification that includes all the operators that currently exist. Relevant for an inventory of knowledge is that a classification of all the operators also offers a framework for mapping all their properties. Additionally, operators of different kinds can be used as a starting point for mapping systems consisting of interacting operators.

Systems of interacting operators. Apart from entities that are an operator, the world is filled with entities that consist of interacting operators. The operator theory refers to every such entity as an 'interaction system'. A distinction is made between two different kinds of interaction systems: 'compound objects' and 'groups'. A compound object consists of operators and/or compound objects that are more strongly materially attached to each other than to their environment. The indication 'more strongly' implies that the separation between the two classes is gradual, and that on the edge intermediates may exist. Yet it is considered practical to work with this division, because most objects can with little doubt be classified in one of these two groups. Examples of compound objects are: a stone, a car, a lump of dirt, a piece of cloth, a planet, a lump of ice in a bucket with water, etc. In contrast to compound objects, a group consists of elements (operators or compound objects) that are not attached materially, e.g. a football team, a school of fish, a colony of bees, a solar system, a gas, the universe, etc.

A mapping of knowledge. Above I have subdivided the territory of knowledge about things in the world into two large areas: operators and interaction systems. For each area I will create a mapping of major fields of knowledge. I also explore in which direction knowledge can potentially be extended.

Knowledge about individual operators. Knowledge about any specific operator involves its overall functioning and construction, and the functioning and construction of its parts. Operators at the bottom of the operator hierarchy (those of the standard model) have a rather basic organization, and their properties are correspondingly few. Higher up in the hierarchy, more complex operators reside, such as animals with brains, which have a very complex structure and which have many properties that frequently can change dramatically over time, e.g. during ageing or during the metamorphosis from egg to butterfly. The operators of different kinds are studied by scientists of different disciplines, e.g. nuclear physicists, chemists, biologists, ethologists, etc. Limits to what can be known about any specific operator are determined by the number of parts that such an operator consists of, the interactions between these parts, and the interactions of the operator with entities in its environment.

Knowledge about interaction systems. When operators interact, the operator theory calls the result an 'interaction system'. The more operators of different kinds are involved, the more interactions are possible. The limit to what can be known about an interaction system will in the end depend on the numbers and kinds of operators involved, and the interactive relationships between these operators. Arto Annila's fundamental viewpoint on this matter is to say that ultimately every operator consists of interacting fundamental particles, and that therefore all the properties of any entity in the universe depend on fundamental particles and their relationships. Within the larger context of Annila's generalization I use the operators as entities that are defined at higher levels of organization, with typical holistic properties as the result of dual closure. In my perspective, the operator theory facilitates the mapping of our knowledge through its stepwise integration of interactions between fundamental particles, resulting in units of higher and higher complexity.

The path to knowledge. Physical cues, such as the cosmic background radiation, have been used to calculate that the universe exists already for 13.799 ± 0.021 billion years. During most of these years humans did not exist. When finally humans evolved and gained the technical and theoretical capacities to investigate their environment, these investigations necessarily started with the 'low-hanging fruit' represented by large things, such as rocks, water, plants and animals. Subsequently, the development of increasingly advanced experimental equipment set the stage for ever more detailed investigations of the world. The first microscope built by Van Leeuwenhoek was instrumental to the discovery of 'animalculi': the unicellular organisms. Later still, chemists discovered molecules, and physicists discovered atoms. Further technical improvements allowed the discovery of the atom nucleus, the protons and neutrons in it, and how protons and neutrons consist of quarks. Every time that scientists discover smaller operators, such discoveries extend the operator hierarchy downward, and at the same time add a 'deeper' layer to the inside of every high-level operator.

Downward limits to knowledge. Around 300 BC, the Greek philosopher Democritus reasoned that when dividing a thing in two parts, at some level undividable objects would be reached: the atoms. In recent decades, scientists have 'cut' matter in parts to discover a long series of objects that decrease in size from molecules, to atoms, to the atom nucleus, to protons and neutrons, and finally to quarks. This ongoing process is frequently used as an indication that there may come no end to the discovery of smaller entities.

Yet, I see two indications of why Democritus's reasoning still deserves our attention. One indication is that one needs unfathomable amounts of energy for the production of things smaller than quarks. A second indication is that if one follows the operator hierarchy downward, the complexity of the dual closures reduces with every step (Figure 21.2). Finally, one arrives at the level of the quarks and other fundamental particles. According to the operator theory quarks have only one closure dimension left; their interface. If one would go one step further down the 'ladder', the theory predicts that one would lose this last closure, and that accordingly one can no longer speak of organized matter. Both lines of reasoning suggest that fundamental particles may well be the lowest level of organization, and that scientific knowledge has a lower limit.

Extending knowledge with every future, higher-level operator. While the 'bottom' of the operator hierarchy suggests the existence of a lowest level of organization, the 'top' of the hierarchy extends in principle in an infinite way (Figure 21.2). This implies that increasingly complex operators can form in the future. It seems therefore that, at least in theory, the complexity of future operators can extend forever.

A thing that may put an end to the formation of increasingly complex operators is the accelerating expansion of the universe. On the long run, accelerating expansion will disperse all matter across a cold and endless space. At some point in the far future, such expansion may block the formation of higher-level operators.

Is knowledge fractal? Anyone who wants to measure the length of the coast of England faces a problem: the result depends on the size of one's yardstick. If one uses a yardstick of one kilometer, the coast will seem shorter

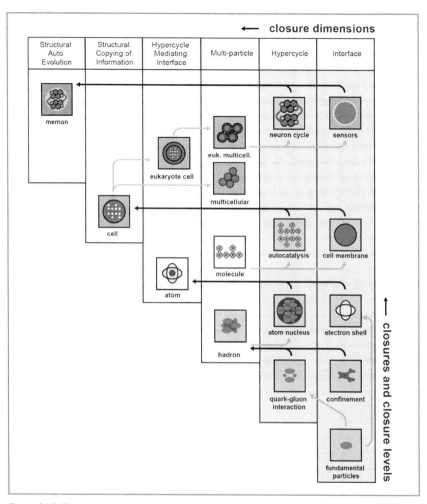

Figure 21.2. The operator hierarchy. Arrows indicate steps towards more complicated organization. Black arrows (pointing to the left) focus on the formation of a new kind of system that is the first of a layer. Gray arrows (pointing to the right) indicate the steps within a layer. Gray columns indicate system organization that represents single closure, either the structural closure of the interface, or the functional closure of a hypercyclic process. The 4 columns on the left indicate operators. Each of these columns indicates a different kind of dual closure, e.g. that of multi-particle, of hypercycle mediating interface, etc. Offering a more detailed explanation of the figure lies beyond the scope of this book. Further information can be found in e.g. Jagers op Akkerhuis (2016). We thank Bentham for permission to reproduce Figure 21.2 after Figure 6 in Chatterjee (2016).

than if one uses a yardstick of one millimeter. The shorter the yardstick, the longer the coast. A similar effect may be at work in science, where increasingly many new facts, of ever greater detail, are discovered if one zooms in on phenomena. While zooming in, however, one simultaneously shifts from complex operators to less complex operators. For example, investigations of the human body may start with limbs and organs, proceed towards tissues and cells, and continue with molecules, atoms and sub-atomic particles. Every step towards a less complex operators can be compared with the use of a smaller yardstick. With each step, the amount of knowledge will increase. Most of such knowledge will, however, be linked to the use of a small yardstick, and may not always be relevant for processes at higher levels. In fact, higher-level closures will generally shield lower level processes from having effects at higher levels.

Conclusion. The operator theory suggests that the operators, and the interaction systems they create, set limits to what can exist in the universe. On this basis one can create a mapping of areas of knowledge. Assuming validity of the operator hierarchy, going down its 'ladder' leads to a decrease in closure dimensions, until a lowest level of organization is reached, representing a lower limit to knowledge. While expansion of the universe and/or instability of future systems may cause practical limits, it is hard to imagine any in principle reason for an upper limit to the complexity of future operators and related interaction systems.

Further reading

Annila, A. (2016) Natural thermodynamics. Physica A: Statistical Mechanics and its Application 444: 843-852.

Chatterjee, A. (2016). Energy, entropy and complexity: thermodynamic and information-theoretic perspectives on ageing. In: M. Kyriazis (ed.) Challenging aging: the anti-senescence effects of hormesis, environmental enrichment, and information exposure. Frontiers in Aging Sciences, Vol. 1, pp. 169-200.

Jagers op Akkerhuis, G.A.J.M. ed. (2016). Evolution and transitions in complexity. The science of hierarchical organization in nature. Springer, Cham, Switzerland.

22. General reflections

Why would anyone explore the possibility of a meta-level theory about hierarchy in the organization of nature in a time when definitions, meta-narratives and essential properties are approached with skepticism and relativism? My drive to work in this direction is linked to my mission and vision as a scientist.

Beyond post-modernism. Of course, and in line with post-modern thinking, I am well aware that descriptions of the things we mean to recognize in nature are relative in the sense that they depend on human observation. And it is only natural that definitions depend on the context in which we use them, which context includes personal goals. In addition, it is easy to see that many concepts we talk about lack natural limits, such as love, the color we call 'green', the border of Wales, a citizen, etc. Without natural limits definitions of such things depend on willful human decisions. All these examples indicate the relativity of many aspects of our knowledge.

It is relevant in this context that the criteria of the operator theory are less relative, because they are chosen in such a way that they allow binary distinctions. Binary distinctions are possible because the criteria that are used overlap with major aspects of natural organization.

The binary nature of the criteria results from topological considerations. As an example, one can think of a piece of rope. This can be a loose piece of rope, or a piece of which the ends are knotted together to construct a closed, circular 'loop' of rope. The 'loop' and the associated criterion of 'being closed' are special, because a system either has such loop, and has closure, or does not have such a loop, and does not have closure. The loop and closure are either present or absent.

The organization of natural systems thus provides us *in special cases* with configurations that allow binary categorization. This allows one to, amidst of many gradual non-binary processes, identify system states that are closed, and to arrange these states in a system with fixed and stringent classes.

Of course not all closures are the same. For example, one can think of the grand cycles ecologists recognize in nature, such as the water cycle, and the carbon cycle. Or one can focus on cyclical processes in the recycling of waste. These are all very important cyclical phenomena that, however, differ in several ways from the dual closure in the operator theory. Firstly, the water cycle, carbon cycle, and waste cycle, do not occur in interaction with a mediating layer, a 'container', or 'interface'. Secondly, in the carbon cycle, carbon is part of many different kinds of objects, and these may take part in different cycles. These examples demonstrate that not every loop in ecosystems or elsewhere is circular in the same way.

Of all possibilities, the operator theory focuses on a combination of two special kinds of closure, called functional closure and structural closure. When they occur together, the functional closure and structural closure are named dual closure (see SCIENCEBITES 1).

Closure combines a causal mechanism and a logical criterion. In nature, dual closures are causally responsible for the formation of systems such as atoms, cells, brains, etc. Because of dual closure an operator can function as a 'whole' singular unity. For example, an atom either has one or more electrons in its shell, or it has none. Without the electrons, the resulting system is not an atom, but a nuclide. The combination of a nucleus and one or more electrons is an essential aspect of what defines an atom. And if a cell loses its dual closure, for example when the closure of the membrane fails, the membrane no longer keeps the catalytic molecules together, and the cell must die. Dual closure thus creates a link between process and classification. Dual closure describes a circular mechanism or shape, while the presence/absence of dual closure allows for a binary logic. This is the reason why I used this combination as the foundation for a theory that integrates the observation of mechanisms that cause specific kinds of organization in nature with a stringent classification.

The criteria for entities with closure can be made very restrictive through the combination of two or more closures. Such combinations at the same time cause, identify and define entities at a specific level of organization, and distinguish these entities from those at other levels of organization. Instead of imposing closure on the system, the system does create the closed configuration by itself, after which this configuration can be recognized by an observer.

Beauty. The use of closures offers a new perspective, that allows one to identify a large group of systems in nature that can be described using well-defined criteria. I think there is beauty in such possibility. To understand what I mean with 'beauty' here, I must explain my mission and vision, and how these have helped me wander through a landscape of post-modern thinking, while searching for a stringent, closure-based ranking that establishes a one-to-one relationship between physical objects and conceptual classes.

Mission. My mission is to develop a theory that contributes to a bottom-up mechanistic reconstruction of how the universe came to be. Such a reconstructive model is based on knowledge gained from observations of current systems, and from the study of remnants of past system states. Examples of such past states are the temperature of the background radiation, and the presence of fossil remains of animals in ancient rocks. Mechanistic insight into the behavior of current systems also provides scientists with a basis for predictions. While working towards a reconstructive model, I follow two major conceptual paths:

Start with what is accessible and expand from there. Historically, scientists always start with picking the 'low-hanging fruit', the easy/local examples, and limited/local theory. Every time that a new aspect of organization is discovered in nature, scientists will attempt to include this in an extended theory (Figure 22.1).

Following the above approach, people expand the explanatory power of scientific knowledge in two directions: (1) they discover or construct new things and processes (new organisms, new forces, new metals, new machines, new culture, etc.); and (2) they construct more general theory (e.g. from Newton to Einstein, from physics to quantum physics, from the history of a country to the history of the universe).

Work towards simplicity, scalability and generality. I aim to avoid unnecessary complexity. I value simplicity, because if the core of a theory is simple yet complete, one will find that the outcome is more scalable and more general than other approaches. In the context of simplicity, it pays to develop stringent definitions that are as general as possible, because it is conceptually

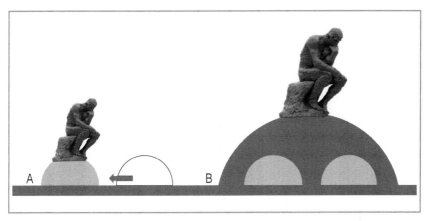

Figure 22.1. From local to general theory. A: a small part of the world is understood (grey blob) and a new scientific area becomes visible (white blob). B: the new scientific area is integrated over time into the theory. Now a new unknown area may be discovered that needs to be integrated.

much easier to work with a stringent and general definition than to constantly have to deal with the exceptions of non-stringent and/or local definitions.

Vision. In my vision 'matter' (in the form of waves and/or particles) provides the basis for observations in science (Figure 22.2). Other aspects can be derived from a material basis.

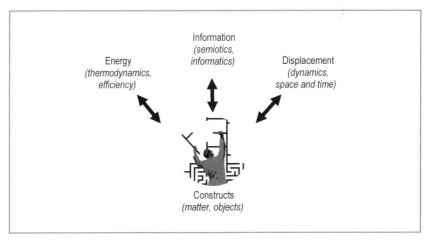

Figure 22.2. Viewing the material world as a foundation for science.

Physical phenomena form the basis of science, the 'facts'. Theory is the logical connection between such facts. The latter indicates that logic, in the form of theory, has a necessary role to play in science. While there may not exist a single 'theory of everything', scientists continue developing interlocking grand theories that in combination offer an integrating view on the many aspects of reality.

Printed in the United States
by Baker & Taylor Publisher Services